Environmentally Sustainable Catalytic Asymmetric Oxidations

Environmentally Sustainable Catalytic Asymmetric Oxidations

Konstantin Bryliakov

CRC Press
Taylor & Francis Group
Boca Raton London New York

CRC Press is an imprint of the
Taylor & Francis Group, an **informa** business

First published in paperback 2024

First published 2015
by CRC Press
2385 NW Executive Center Drive, Suite 320, Boca Raton FL 33431

and by CRC Press
4 Park Square, Milton Park, Abingdon, Oxon, OX14 4RN

First issued in hardback 2019

CRC Press is an imprint of Taylor & Francis Group, LLC

© 2015, 2019, 2024 Taylor & Francis Group, LLC

Library of Congress Cataloging-in-Publication Data

Bryliakov, Konstantin.
  Environmentally sustainable catalytic asymmetric oxidations / Konstantin Bryliakov.
     pages cm
  "A CRC title."
  Includes bibliographical references and index.
  ISBN 978-1-4665-8857-8
  1. Oxidation. 2. Green chemistry. 3. Chemistry, Organic. I. Title.

QD281.O9B79 2015
547'.23--dc23   2014022375

ISBN: 978-1-4665-8857-8 (hbk)
ISBN: 978-1-03-292950-7 (pbk)
ISBN: 978-0-429-16826-0 (ebk)

DOI: 10.1201/b17422

Visit the Taylor & Francis Web site at
http://www.taylorandfrancis.com

and the CRC Press Web site at
http://www.crcpress.com

Contents

Preface

The discovery of simple and efficient catalyst systems for the asymmetric oxidation of organic compounds is an important and challenging task of modern chemistry. Synthetic chiral compounds are represented in various areas of everyday life: for example, many top worldwide pharmaceuticals [1] are chiral substances of synthetic origin. Today, the demand for novel, efficient, and sustainable regio- and stereoselective processes for the oxidation of organic substrates continues to grow in line with toughening economic and environmental constraints.

Among the types of asymmetric processes, catalytic transformations that are capable of creating many molecules of chiral products per single chiral catalyst molecule are the most challenging and valued, as recognized by the 2001 Nobel Prize awarded to W. S. Knowles, R. Noyori, and K. B. Sharpless to acknowledge their contributions to chirally catalyzed hydrogenation and oxidation reactions.

This book deals with the latter types of asymmetric transformations, covering existing transition metal-based and ecologically friendly catalyst systems (essentially those using H_2O_2 or O_2 as the ultimate oxygen source) for various asymmetric oxidation processes such as epoxidations, sulfoxidations, cis-dihydroxylations, Baeyer–Villiger oxidations, kinetic resolution of secondary alcohols, oxidative coupling of 2-naphthols, and others (Chapters 2 through 4). In addition to transition metal-based catalyst systems, organocatalytic systems are also reviewed (Chapter 5).

Special attention is given to stereospecific C–H oxidations with H_2O_2. The problem of selective catalytic oxidation of non-activated C–H groups of hydrocarbons under mild conditions has been successfully solved by nature in the course of biological evolution (through the use of metalloenzymes). Synthetic catalysts reported to date do not exhibit comparable efficiencies and selectivities. A biomimetic approach of mimicking the reactivities of natural metalloenzymes with synthetic metallocomplexes is the most logical way to bridge this gap and provide a basis for devising ecologically friendly stereoselective oxidations of C–H groups. Chapters 6 and 7 are dedicated to Fe- and Mn-based biomimetic catalyst systems.

The author hopes this work can serve as a reference book for both academic and industrial researchers as well as university faculty and students at graduate and postgraduate levels.

Konstantin Bryliakov

Author

Konstantin P. Bryliakov graduated from Novosibirsk State University in 1999. He earned a Cand. Chem. Sci. (PhD) in chemical physics from the Institute of Chemical Kinetics and Combustion (Novosibirsk) in 2001 under the direction of Professor E. P. Talsi. In 2008, Dr. Bryliakov was awarded a Dr. Chem. Sci. in catalysis from the Boreskov Institute of Catalysis (Novosibirsk).

Dr. Bryliakov is a leading research scientist at the Boreskov Institute. He has co-authored more than 90 papers, book chapters, and patents. His research interests include transition metal-catalyzed asymmetric oxidations, single-site olefin polymerizations, and mechanistic aspects of those transformations.

List of Abbreviations

Ac	acetyl
AE	asymmetric epoxidation
A:K	alcohol:ketone ratio
Binol	1,1′-bi-2-naphthol
Bn	benzyl
Boc	*t*-butyloxycarbonyl
cAE	catalytic asymmetric epoxidation
CLAMPS	cross-linked aminomethylpolystyrene
DBU	1,8-diazabicyclo[5.4.0]undec-7-ene
DIC	diisopropylcarbodiimide
DMAP	*N,N*-dimethyl-4-aminopyridine
DMCH	1,2-dimethylcyclohexane
ee	enantiomeric excess
EPR	electronic paramagnetic resonance
H$_2$dipic	2,6-pyridinedicarboxylic (dipicolynic) acid
IBA	isobutyric aldehyde
IBAc	isobutyric acid
MA	maleic anhydride
MS	molecular sieves
MTBE	methyl *tert*-butyl ether
MTO	methyltrioxorhenium
Naph	naphthyl
N-Me-Imd	*N*-methylimidazole
6-Me$_3$-tpa	tris(6-methyl-2-pyridylmethyl)amine
NMO	*N*-methylmorpholine-*N*-oxide
NMR	nuclear magnetic resonance
N-Oct-Imd	*N*-octylimidazole
OTf	trifluoromethanesulfonate (triflate) anion
PA	pivalaldehyde
PEG	polyethylene glycol
Ps	polystyrene
Py-*N*-O	pyridine-N-oxide
RC	retention of stereoconfiguration
Salan	*N,N′-o*-(hydroxybenzyl)-1,2-diaminoethane
Salen	*N,N′*-(salicylidene)-1,2-ethylenediamine
S-TBNBr	SBA-16/*N*-propyl-*N,N,N*-tri-*n*-butylammonium bromide
TBHP	*tert*-butylhydroperoxide
TEA	triethylamine
TEMPO	2,2,6,6-tetramethylpiperidinooxyl
TFA	trifluoroacetic acid
TOF	turnover frequency

TON   turnover number
TPP   tetraphenylporphyrin
Triflate (OTf⁻) trifluoromethanesulfonate ((CF$_3$)O$_2$SO⁻)
TTM   time to market
UHP   urea hydroperoxide

1 Introduction

Chirality, as a natural phenomenon, has been a crucial factor for biological life and evolution because many metabolic processes involve chiral reactants and catalysts and thus rely on the processes of asymmetric catalysis and chiral recognition. Tremendous efforts of chemists and biologists have been invested into understanding, reproducing, and prototyping these naturally existing phenomena to find artificial pathways for creating valuable chiral molecules that possess biological activities.

The demand for bioactive chiral compounds grows continuously because chiral building blocks are indispensable for synthesizing biologically active compounds such as pharmaceuticals, agrochemicals, flavors, fragrances, and advanced material designs. The four basic approaches to accessing enantiomerically pure chiral organic compounds are [2]:

Separation of enantiomers via resolution of racemic mixtures—This is classical resolution via diastereomers, chromatographic separation, enzymic resolution, chemical kinetic resolution, and other techniques. Major disadvantages of this approach are the need for large amounts of solvents and the loss (or necessity to recycle) of undesirable stereoisomers (ca. 50% of material).

Chiral pool approach—This use of chiral building blocks originating from natural products for the construction of a target molecule is very convenient because naturally occurring materials often have high enantiomeric purity levels. However, from an industrial perspective, commercial availability of the starting material may present a substantial limitation.

Enzymic and microbial (biocatalytic) asymmetric transformations—Although they usually proceed with high stereoselectivity, biocatalytic transformations have specific disadvantages, for example, (a) development of an efficient biocatalyst can take a long time and (b) isolation of the product from dilute aqueous solutions may be tedious and expensive.

Asymmetric synthesis—This synthesis of organic molecules for generating asymmetric induction in the course of chemical transformations of achiral (pro-chiral) substrates is the main subject of this book. In particular, the discussion focuses on asymmetric synthesis in the presence of chiral catalysts. This is probably the most advantageous strategy since a single chiral catalyst molecule can (at least in theory) create millions of molecules of the chiral products—as enzymes do in biological systems. In recent years, a number of comprehensive monographs and edited collections dedicated to various aspects of catalyzed asymmetric synthesis may be useful for interested readers [2–25].

The scope of this book is limited to catalytic asymmetric oxidation reactions. Catalyzed selective oxidations constitute an important class of transformations that have been widely exploited by the chemical industry. A shift to more sustainable and ecologically friendly (greener) catalytic oxidation processes reflecting increasing

1

community environmental concerns has become more apparent both in the laboratory and on an industrial scale [26–31]. Apparently, such green chemistry issues as reduction (or avoidance) of the use of hazardous substances, energy, and wastes are of particular importance for fine chemical and pharmaceutical industries that directly impact the health and environmental aspects of human life [31–39].

For green chemistry, catalysis is a phenomenon of special significance because catalyzed processes may minimize the quantities of reagents, energy, and wastes (including wastes from the resolution of racemates). In the asymmetric oxidation area, this can be achieved by using highly active, regioselective and stereoselective, and robust catalysts (preferably recyclable, non-toxic, metal-based, or metal-free) and minimum amounts of (preferably recyclable) solvents and non-hazardous oxidants. This leaves much space for further progress in the field.

An ideal catalytic asymmetric transformation should proceed with 100% conversion and provide complete chemo-, regio-, and stereo-control, and allow recycling of catalysts and reaction solvents. For large-scale industrial applications, catalysts should be recoverable and non-toxic, withhold high turnover numbers, exhibit high turnover frequencies, and require minimum solvents and additives. The overall process should produce minimum waste (which implies high selectivity) and be inexpensive and easy to implement (see Chapter 8). To date, few catalyst systems meet these rigorous requirements.

The major requirements for catalytic asymmetric oxidation methodologies are (a) high chemoselectivity and stereoselectivity (to maximize the yield of the desired product and minimize waste), (b) high efficiency as evidenced by turnover number and turnover frequency (to minimize catalyst load and achieve transformation within a reasonably short time), and (c) the ability to use environmentally benign oxidants. Indeed, while a solvent may be in most cases distilled and recycled, minimization of the wastes resulting from transformation of stoichiometric oxidants in the course of an enantioselective process is crucial for achieving overall sustainability.

Therefore, chlorine-containing and high molecular weight oxidants (like sodium hypochlorite, iodosylarenes, alkyl hydroperoxides, most peroxycarboxylic acids, and oxone) should be avoided. They are either harmful or generate large quantities of wastes thus leaving cheap and non-polluting oxidants such as hydrogen peroxide and dioxygen as the protagonists of sustainable catalyst systems discussed later. Both oxidants are readily available and contain similar amounts of active oxygen (47 and 50%, respectively) [40]. The "green" status of hydrogen peroxide seems more favorable since it is safer for work with organic solvents and produces water as its only by-product. Molecular oxygen often requires (at least) a stoichiometric amount of organic co-reductant that leads to generation of organic by-products.

An industrially important issue is the amount of time required to take a new chemical process from laboratory scale to the market. In the current economy in which important issues such as time to market and revenue remain in a more or less inverse proportional relationship, time-proven technologies are preferred often over more advanced and environmentally friendly processes that require time and capital investment for development and implementation. For that reason, the needs of the fine chemical and pharmaceutical industries to develop novel green asymmetric catalytic processes with broad scope and utility that may be adapted for various processes quickly are of major interest.

In this book we discuss transition-metal catalyzed epoxidations, sulfoxidations, *cis*-dihydroxylations, Baeyer–Villiger oxidations, oxidative kinetic resolution of secondary alcohols and desymmetrization of *meso*-diols, oxidative coupling of 2-naphthols, and others. The nature of active species and the reaction mechanisms are mentioned as needed within the text. The rapidly emerging area of organocatalytic asymmetric oxidations is discussed too.

Chapters 6 and 7 are dedicated to Fe- and Mn-based biomimetic catalyst systems for the stereospecific oxidation of C–H functional groups that may serve as bases for future catalyst systems for stereoselective C–H oxidations. Existing and potential practical applications of sustainable catalytic asymmetric oxidations are mentioned briefly where appropriate. At the end of each chapter, a figure illustrating a few preparative scale reactions (if applicable) is provided.

REFERENCES

1. http://www.pharmacytimes.com/publications/issue/2013/July2013/Top-200-Drugs-of-2012
2. Blaser, H. U., Spindler, F., and Studer, M. 2001. Enantioselective catalysis in fine chemical production. *Appl. Cat. A: General* 221: 119–143.
3. Noyori, R. 1994. *Asymmetric Catalysis in Organic Synthesis*. New York: John Wiley & Sons.
4. Jacobsen, E. N., Pfalz, A., and Yamamoto, H., Eds. 1999. *Comprehensive Asymmetric Catalysis*. Berlin: Springer.
5. Ojima, I., Ed. 2000. *Catalytic Asymmetric Synthesis*, 2nd ed. New York: John Wiley & Sons.
6. Bäckvall, J. E., Ed. 2004. *Modern Oxidation Methods*. Weinheim: Wiley-VCH.
7. Blaser, H. U., Schmidt, E., eds. 2004. *Asymmetric Catalysis on Industrial Scale*. Weinheim: Wiley-VCH.
8. Malhotra, S. V., Ed. 2004. *Methodologies in Asymmetric Catalysis*. Washington, DC: American Chemical Society.
9. Berkessel, A. and Groger, H. 2005. *Asymmetric Organocatalysis: From Biomimetic Concepts to Applications in Asymmetric Synthesis*. Weinheim: Wiley-VCH.
10. Mikami, K. and Lautens, M., Eds. 2007. *New Frontiers in Asymmetric Catalysis*. Hoboken: Wiley-Interscience.
11. Enders, D. and Kaeger, K. E., Eds. 2007. *Asymmetric Synthesis with Chemical and Biological Methods*. Weinheim: Wiley-VCH.
12. Dalko, P. I. 2007. *Enantioselective Organocatalysis*. Weinheim: Wiley-VCH.
13. Ding, K. and Uozumi, Y., Eds. 2008. *Handbook of Asymmetric Heterogeneous Catalysis*. Weinheim: Wiley-VCH.
14. Caprio, V. and Williams, J. 2009. *Catalysis in Asymmetric Synthesis*, 2nd ed. London: Wiley-Blackwell.
15. Walsh, P. J. and Kozlowski, M. C. 2009. *Fundamentals of Asymmetric Catalysis*. Sausalito: University Science Books.
16. Ojima, I., Ed. 2010. *Catalytic Asymmetric Synthesis*, 3rd ed. Hoboken: Wiley-Interscience.
17. List, B., Ed. 2010. *Asymmetric Organocatalysis*. Dordrecht: Springer.
18. Pellisier, H. 2010. *Recent Developments in Asymmetric Organocatalysis*. Cambridge: Royal Society of Chemistry.

19. Patti, A. 2011. *Green Approaches to Asymmetric Catalytic Synthesis*. Heidelberg: Springer.
20. Gruttadauria, M. and Giacalone, F., Eds. 2011. *Catalytic Methods in Asymmetric Synthesis: Advanced Materials, Techniques, and Applications*. Hoboken: John Wiley & Sons.
21. Mahrwald, R., Ed. 2011. *Enantioselective Organocatalyzed Reactions I*. Dordrecht: Springer.
22. Mahrwald, R., Ed. 2011. *Enantioselective Organocatalyzed Reactions II*. Dordrecht: Springer.
23. Ma, S., Ed. 2011. *Asymmetric Catalysis from a Chinese Perspective*. Dordrecht: Springer.
24. Šebesta, R., Ed. 2012. *Enantioselective Homogeneous Supported Catalysis*. Cambridge: Royal Society of Chemistry.
25. Bryliakov, K. P. 2013. Sustainable asymmetric oxidations. In *Comprehensive Inorganic Chemistry II*, Reedijk, J. and Poeppelmeier, K., Eds. Oxford: Elsevier, pp. 625–664.
26. Anastas, P. T. and Warner, J. C. 1998. *Green Chemistry: Theory and Practice*. New York: Oxford University Press.
27. Anastas, P. T. and Kirchhoff, M. M. 2002. Origins, current status, and future challenges of green chemistry. *Acc. Chem. Res.* 35: 686–694.
28. Cavani, F. and Teles, J. H. 2009. Sustainability in catalytic oxidation: an alternative approach or a structural evolution? *ChemSusChem* 2: 508–534.
29. Hermans, I., Spier, E. S., Neuenschwander, U. et al. 2009. Selective oxidation catalysis: opportunities and challenges. *Top. Catal.* 52: 1162–1174.
30. Tucker, J. L. 2010. Green chemistry: cresting a summit toward sustainability. *Org. Proc. Res. Dev.* 14: 328–331.
31. Sheldon, R. A. 2012. Fundamentals of green chemistry: efficiency in reaction design. *Chem. Soc. Rev.* 41: 1437–1451.
32. Iida, T. and Mase, T. 2002. Scalable enantioselective processes for chiral pharmaceutical intermediates. *Curr. Opin. Drug Discovery Dev.* 5: 834–851.
33. Luo, S. Z., Peng, Y. Y., Zhang, B. L. et al. 2004. Some new trends and recent progress toward environmentally benign synthesis. *Curr. Org. Synth.* 1: 405–429.
34. Federsel, H. J. 2005. Asymmetry on large scale: the roadmap to stereoselective processes. *Nat. Rev. Drug Discovery* 4: 685–697.
35. Constable, D. J. C., Dunn, P. J., Hayler, J. D. et al. 2007. Key green chemistry research areas: a perspective from pharmaceutical manufacturers. *Green Chem.* 9: 411–420.
36. Walsh, P. J., Li, H. M., and de Parrodi, C. A. 2007. A green chemistry approach to asymmetric catalysis: solvent-free and highly concentrated reactions. *Chem. Rev.* 107: 2503–2545.
37. Cue, B. W. and Zhang, J. 2009. Green process chemistry in the pharmaceutical industry. *Green Chem. Lett. Rev.* 2: 193–211.
38. Andrews, I., Cui, J., DaSilva, J. et al. 2010. Green chemistry articles of interest to the pharamaceutical industry. *Org. Proc. Res. Dev.* 14: 19–29.
39. Busacca, C. A., Fandrick, D. R., Song, J. J. et al. 2011. Growing impact of catalysis in the pharmaceutical industry. *Adv. Synth. Catal.* 353: 1825–1864.
40. Strukul, G. and Scarso, A. 2013. In *Liquid Phase Oxidation via Heterogeneous Catalysis*. Clerici, M. G. and Kholdeeva, O. A., Eds. Hoboken: John Wiley & Sons, pp. 1–20.

2 Transition Metal-Catalyzed Asymmetric Epoxidations

The significance of catalytic asymmetric epoxidation (cAE) for synthetic chemistry stems from the relative chemical inertness of olefinic groups that require chemical functionalization before use in fine chemical synthesis. Catalytic asymmetric epoxidation of olefins is a convenient methodology leading to versatile and reactive yet stable intermediates containing one or two stereogenic centers that can be involved in further transformations via asymmetric ring-opening reactions [1].

R_1, R_2 = alkyl, aryl

In 1976, Wynberg and co-workers pioneered the catalytic synthesis of epoxides of α,β-unsaturated ketones with up to 25% *ee* in the presence of cinchona alkaloid-derived quaternary ammonium chloride salt [2]. In 1977, two research groups independently demonstrated the possibility of enantioselective epoxidation of allylic alcohols with alkyl hydroperoxides in the presence of chiral molybdenum [3] and vanadium [4] complexes, but the enantioselectivities were only moderate (<44% *ee*).

Two years later, the first reports on the cAEs of unfunctionalized olefins by *t*-butyl hydroperoxide (TBHP) in the presence of chiral molybdenum catalysts appeared, documenting low-to-moderate enantioselectivities (<35% *ee*) [5,6]. In 1980, Katsuki and Sharpless reported the first "practical" example of catalytic asymmetric epoxidation. Several allylic alcohols were oxidized with TBHP in good yields and at >95% *ee* in the presence of a titanium complex with an optically active tartaric acid ester [7].

After 1980, many catalyst systems appeared, relying on either transition metal-based or purely organic catalysts. The best known are Katsuki's [8–11] and Jacobsen's [12–15] manganese–salen based systems for the epoxidation of unfunctionalized olefins. Interested readers may find recent critical reviews discussing catalytic asymmetric epoxidation with various oxidants [16–30] useful. This chapter provides an up-to-date survey of existing transition metal-based catalyst systems for asymmetric epoxidations (AEs) with O_2 and H_2O_2.

MANGANESE SYSTEMS

MANGANESE–SALEN AND RELATED SYSTEMS USING O_2 AS TERMINAL OXIDANT

The first documented attempt to use molecular oxygen as an oxidant for cAE was reported in 1992 by Mukaiyama and co-workers who studied enantioselective oxidation in the presence of chiral manganese–salen complexes of types **1a** and **1b** (R = Me, Figure 2.1); enantioselectivities of 43 to 77% *ee* were reported [31]. Catalysts of those types operated as monooxygenases, and the catalyst systems required the use of a co-reductant (pivalaldehyde, 3 equivalents). Even higher enantioselectivities in the epoxidation of 2,2-dialkylchromenes were demonstrated by Jacobsen's catalyst **1c** (up to 92% ee), but with low chemical yields (24 to 37%) [32].

Structurally related chiral β-ketoiminato manganese complexes of type **2** catalyzed the epoxidation of conjugated olefins with moderate to good enantioselectivities (33 to 84% ee) and 22 to 70% chemical yield [33–36]. Disadvantages of Mukaiyama-type systems were the high catalyst loads (usually 12 mol%) and an excess of an aldehyde co-reductant that contaminated the reaction mixture, affording a threefold excess of the corresponding carboxylic acids.

On the other hand, Mukaiyama's systems demonstrated enviable versatility. In addition to olefins, they could enantioselectively oxidize substituted methyl phenyl sulfides to sulfoxides, demonstrating moderate to high optical yields (24 to 94% ee) and moderate to good yields [37,38]. Besides aliphatic aldehydes, 2-alkyl-2-oxocyclopentanecarboxylates may be used as sacrificial reductants [39].

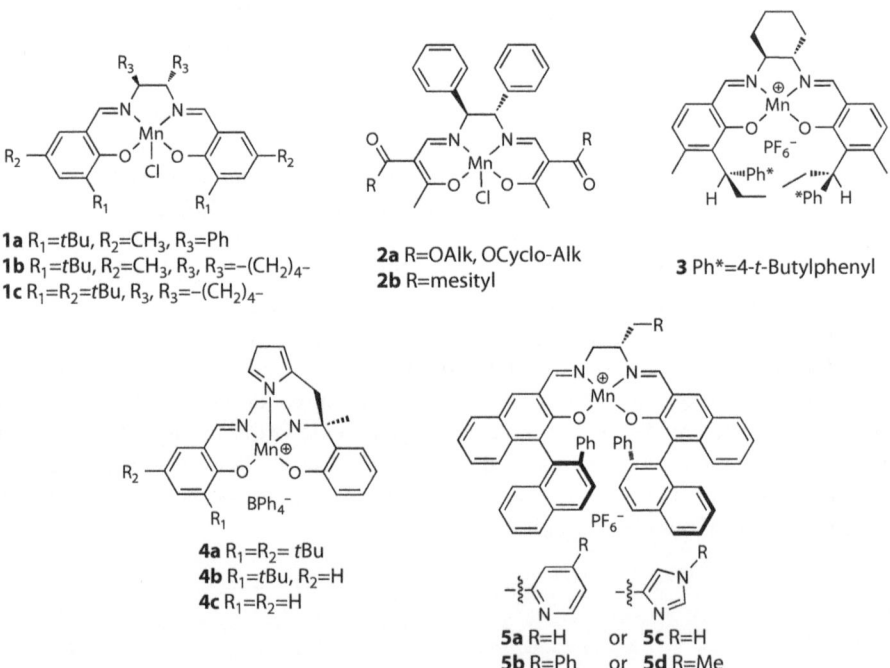

1a R_1=*t*Bu, R_2=CH$_3$, R_3=Ph
1b R_1=*t*Bu, R_2=CH$_3$, R_3=–(CH$_2$)$_4$–
1c R_1=R_2=*t*Bu, R_3, R_3=–(CH$_2$)$_4$–

2a R=OAlk, OCyclo-Alk
2b R=mesityl

3 Ph*=4-*t*-Butylphenyl

4a R_1=R_2= *t*Bu
4b R_1=*t*Bu, R_2=H
4c R_1=R_2=H

5a R=H   or   **5c** R=H
5b R=Ph   or   **5d** R=Me

FIGURE 2.1   Salen– and β-ketoiminato–manganese complexes.

Interestingly, Mukaiyama with co-workers discovered that additives of N-donor organic bases (such as N-alkyllimidazole, 0.5 equivalent) enhanced the enantiomeric ratio and reversed the sense of asymmetric induction [32,36,40,41]. To explain these observations, it was proposed that different intermediates could operate under different reaction conditions. In particular, in the absence of a nitrogen base, a manganese (III)–acylperoxo complex intermediate was expected, while the addition of N-methylimidazole (that assisted O-O bond heterolysis in the initially formed intermediate) converted the acylperoxo complex to an oxomanganese (V) complex [32,41,42].

Lee and co-workers reported an enantioselective aerobic epoxidation of alkenes (in up to 81% *ee* for 2,2-dimethylchromene) catalyzed by Jacobsen's complex **1b**; the actual oxidant was dichlorocarbonyl oxide (formed in situ from added chloroform and sodium hydroxide) [43]. However, the conversion rates observed were rather low (14 to 38% in most cases), and the major reaction products were hem-dichlorocyclopropanes.

Several manganese–salen complexes were tested in electrocatalytic AEs with molecular oxygen; for styrene, moderate *ee* values (up to 67%) and low yields (17 to 19%) were reported [44]. Using 5 mol% of Jacobsen's catalyst **1b**, Tanaka et al. reported moderately to highly enantioselective (30 to 87% *ee*) electrocatalytic epoxidation of various alkenes in higher yields (up to 93%) using a CH_2Cl_2/aqueous NaCl two-phase media [45]. Some results of enantioselective aerobic epoxidations are collected in Table 2.1.

MANGANESE–SALEN AND RELATED SYSTEMS USING H_2O_2 AS TERMINAL OXIDANT

The first successful manganese-catalyzed enantioselective epoxidation of olefins was reported in 1993 by Meunier and co-workers. They epoxidized 4-chlorostyrene with H_2O_2 in the presence of catalyst **1b** (3.6 mol%) with 39% optical yield [46,47]. To increase the epoxidation enantioselectivity, the authors used a donor additive, 4-*tert*-Bu-pyridine [46]. At the same time, enantioselective oxidation of several alkyl aryl sulfides to sulfoxides with H_2O_2 in acetonitrile was reported to proceed with 84 to 95% yield and 34 to 68% *ee* in the presence of manganese(III)–salen complexes [47].

In 1994, Pietikäinen investigated the catalytic activity of chiral manganese–salen complexes of types **1a** and **1c** in the epoxidation of conjugated olefins 1,2-dihydronaph-thalene and E-β-methylstyrene in a dichloromethane/methanol mixture [48]. The author examined the effects of various donor additives: 4-dimethylaminopyridine + benzoic acid, imidazole; N-methylimidazole (most productive based on epoxide yield of 34 to 63% and enantioselectivity of 42 to 60% *ee*); 1,2-dihydronaphthalene was epoxidized with higher enantioselectivity (60% versus 42 to 47% for E-β-methylstyrene).

Katsuki and co-workers epoxidized 2,2-dimethylchromene derivatives on a man-ganese (III) catalyst 3 (Figure 2.1) in the presence of an N-methylimidazole additive (axial ligand) [49]. The authors screened a series of solvents of which acetonitrile was the best. Enantioselectivities of 88 to 95% *ee* were reported along with rather high epoxide yields (up to 98%) [49]. Katsuki's system used a 2 mol% catalyst load

---

TABLE 2.1
Asymmetric Epoxidations with O_2 Catalyzed by Manganese Complexes

N	Substrate	Catalyst	Oxidant	Additives	Epoxide Yield(%)	ee (%)	Ref.
1		**1b**	O_2/PA[a]	N-Me-Imd	52	83	[32]
		1b	O_2/PA[a]	—	57	79	[35]
2		**1b**	O_2/IBA	N-Me-Imd	12	91	[32]
		1c	O_2/	Imidazole[a]	15	81	[43]
3		**1c**	O_2	Imidazole	10	43	[43]
4		**1b**	O_2	Imidazole	19	23	[43]
5		**2b**	O_2/PA[a]	—	18 (cis)	80	[36]
		1c	O_2	Imidazole	6	80	[43]
6		**1b**	O_2/IBA	N-Oct-Imd	34	91	[32]

[a] In the presence of $CHCl_3$, NaOH, and Bu_4NBr.

Note: PA = pivalic anhydride. N-Me-Imd = *N*-methylimidazole. IBA = isobutyric anhydride. N-Oct-Imd = *N*-octylimidazole.

but as much as 10 equivalents of H_2O_2 (versus 2.3 to 4.1 equivalents of H_2O_2 in the protocols of Meunier and Pietikäinen), apparently, due to high catalase activity of complex **3**, leading to degradation of H_2O_2.

The key role of added nitrogen base was discussed in a review [50]. Apparently, imidazole derivatives serve as both proximal ligands and as acid-base catalysts to favor the heterolytic cleavage of the peroxidic O-O bond in the initially formed hydroperoxo metal species to yield a reactive oxometal intermediate [50]. An axial imidazole donor is often encountered in the structures of many peroxidases and is believed to facilitate O-O heterolysis over the O-O homolysis pathway leading to destructive radical reactions [51-55]. To avoid the use of external donors, Berkessel and co-workers designed a series of pentacoordinate salalen complexes of type **4** (Figure 2.1) with an imidazole "arm" covalently bonded to the salen complex [56,57]. Using 10 mol% of catalyst, 1,2-dihydronaphthalene was epoxidized with 1% hydrogen peroxide (10 equivalents) with up to 64 to 66% *ee* [56,57].

More recently, Shitama and Katsuki used a similar approach to synthesize second-generation salen complexes of type **5**, featuring elements of axial chirality at the 3,3' positions of the aldehyde moieties [58,59]. Using 2.5 to 5.0 mol% of catalyst **5** and 3 equivalents of the oxidant (30% H_2O_2), the authors reported the epoxidation of chromene derivatives with H_2O_2 with high yields and *ee* results up to 97 to 99%. The *N*-methylimidazole arm (in complex **5d**) emerged as the most productive in achieving high enantioselectivities [58,59].

Another approach to suppress the formation of hydroxo radicals and the catalase activity of manganese–salen complexes was reported. Pietikäinen used soluble salts (acetates, hydrocarbonates, formates, benzoates) or nitrogen bases as additives and reported the oxidation of various alkenes in a dichloromethane/methanol mixture with aqueous H_2O_2 or urea-H_2O_2 (UHP) in 64 to 96% *ee* in the presence of catalysts of the type **1** (5 mol%) [60]. Ammonium acetate (20 mol%) generally gave better results than heterocyclic nitrogen bases.

Ammonium acetate was also used as additive by Kureshy and co-workers. They successfully epoxidized several conjugated alkenes with UHP over a methylene-bridged dimeric homochiral manganese–salen complex (2.5 mol%); nearly quantitative conversion and 100% *ee* were reported for some substituted 2,2-dimethylchromene derivatives [61,62]. Later, the authors synthesized macrocyclic chiral salen–Mn (II) complexes and studied their activities in enantioselective epoxidation of several unfunctionalized olefins [63].

When discussing the modes of action of carboxylate salt co-catalysts, the possibilities invoked were the formation of reactive peroxycarboxylic species [60,64] and promotion of the formation of manganese hydroperoxo complexes through salt basicity [60]. An alternative approach was explored by Pietikäinen who reported the asymmetric epoxidation over catalysts of the type **1** with peroxycarboxylic acids formed in situ from various solid H_2O_2 adducts and carboxylic acid anhydrides using *N*-methylmorpholine *N*-oxide (NMO) as additive. The author concluded that the combination of UHP, maleic anhydride, and NMO led to pronounced increases in reactivity and enantioselectivity [65].

Extensive research on manganese–salen catalysis of asymmetric epoxidations with H_2O_2 revealed a rather narrow scope of good substrates, mostly limited to substituted 2,2-dimethylchromenes that often epoxidized with *ee* over 90%; other substrates were not as productive. Some results of epoxidations by manganese–salen and H_2O_2 catalysts systems are listed in Table 2.2.

Few enantioselective oxidations of substrates other than unfunctionalized olefins with H_2O_2 over manganese–salen type complexes have been reported. Brun and co-workers epoxidized allylic alcohols (geraniol and nerol) with moderate *ee*s (50 to 55%) without the addition of external bases [66,67]. Unlike the manganese–salen catalyst systems using iodosylarenes and peroxycarboxylic acids as terminal oxidants in which the reactive oxygen transferring species were discussed extensively [21], the nature of active species in similar catalyst systems using H_2O_2 remains poorly explored.

OTHER MANGANESE-BASED SYSTEMS

In the late 1990s, other types of manganese-based catalyst systems for asymmetric oxidation of alkenes with H_2O_2 appeared. Bolm and co-workers synthesized C_3-symmetric chiral 1,4,7-triazacyclononane derivatives of type 6 (Figure 2.2) and tested them as ligands for in situ generation of chiral catalysts by the reaction with $Mn(OAc)_2$ at a 1.5:1.0 ligand-to-Mn ratio [68].

Styrene was epoxidized in methanol with up to 43% ee, using 2 to 3 equivalents of H_2O_2 and 3 mol% Mn load. Interestingly, when epoxidizing *cis*-β-methylstyrene, the authors identified the major product as a *trans*-epoxide (55% *ee*) [68]. This may indicate a stepwise reaction mechanism with intermediate formation of relatively long-lived carbon-centered radical or cationic intermediates [17,50]. Later, several *p*-substituted styrenes were epoxidized with generally low enantioselectivities (15 to 26% *ee*) when using L-proline derived triazacyclononane ligand 7 as the chiral auxiliary; the active catalyst was a dinuclear Mn (III) complex [69].

Catalysts prepared in situ from macrocyclic ligands of the types 8 to 10 with $Mn(OAc)_2$ or $MnSO_4$ also demonstrated low yields and enantioselectivities (1 to 23% *ee*) [70,71].

In 2003, Stack and co-workers reported that manganese (II) complexes with ligands of type 11 (Figure 2.3) and others could catalyze the epoxidations of various alkenes with commercial peracetic acid in nearly quantitative yield [72–74]. Some of the complexes were chiral, but the epoxidation enantioselectivity was not scrutinized. Costas and co-workers synthesized chiral manganese (II) triflate complexes with more sophisticated pinene-derived ligands (structure 12a and its enantiomer) and showed that those could act as chiral inducers in alkane epoxidations with AcOOH (up to 46% *ee*) [75].

Later, several manganese (II) aminopyridine triflate complexes with ligands of the common structure 13 were shown to catalyze the enantioselective epoxidations of olefins with various oxidants (peracetic acid, alkyl hydroperoxides, iodosylarenes) with up to 79% *ee*. Some conclusions about the nature of catalytically active sites responsible for the enantioselective oxygen transfer were drawn [76]. The efficiencies of manganese triflate catalysts with ligands 11, 12, and 13 were remarkable. The latter successfully operated at low catalyst loads of 1.0 to 0.1 mol% [72–76] versus 2 to 10 mol% for manganese–salen catalysts (see above).

An important breakthrough in aminopyridine-based catalyst systems was made in 2009 when Costas and co-workers showed that manganese triflate complexes (with

TABLE 2.2
Asymmetric Epoxidation with H_2O_2 and Related Oxidants Catalyzed by Manganese–Salen Complexes

N	Substrate	Catalyst	Oxidant	Additive	Epoxide Yield (%)	ee (%)	Ref.
1		1c	H_2O_2	4-methylpyridine	Not reported	39	[47]
2		1c	H_2O_2	N-Me-Imd	53	52	[48]
		1a	H_2O_2	N-Me-Imd	59	60	[48]
		1c	UHP/MA	NMO	70	73	[66]
		1a	H_2O_2	NH_4OAc	73	67	[60]
		4a	H_2O_2	—	72	64	[56]
3		1c	UHP/MA	NMO	81	90	[65]
		1c	H_2O_2	NH_4OAc	71	87	[60]
		5d	H_2O_2	—	95	88	[58]
4		3	H_2O_2	N-Me-Imd	98	95	[49]
		dimeric	UHP	NH_4OAc	>99[a]	84	[61]
		5d	H_2O_2	—	80	98	[58]
5		1c	H_2O_2	NH_4OAc	90	91	[60]
		5d	H_2O_2	—	84	98	[59]
6		1c	H_2O_2	NH_4OAc	84	96	[60]
7		1c	H_2O_2	N-Me-Imd	34	47	[48]
		5d	H_2O_2	—	58	31	[58]
8		dimeric	UHP	NH_4OAc	100	100	[61]
9		4b	H_2O_2	—	67	52	[57]

Note: N-Me-Imd = *N*-methylimidazole. UHP = urea hydroperoxide. MA = maleic anhydride. NMO = *N*-methylmorpholine *N*-oxide

[a] alkene conversion

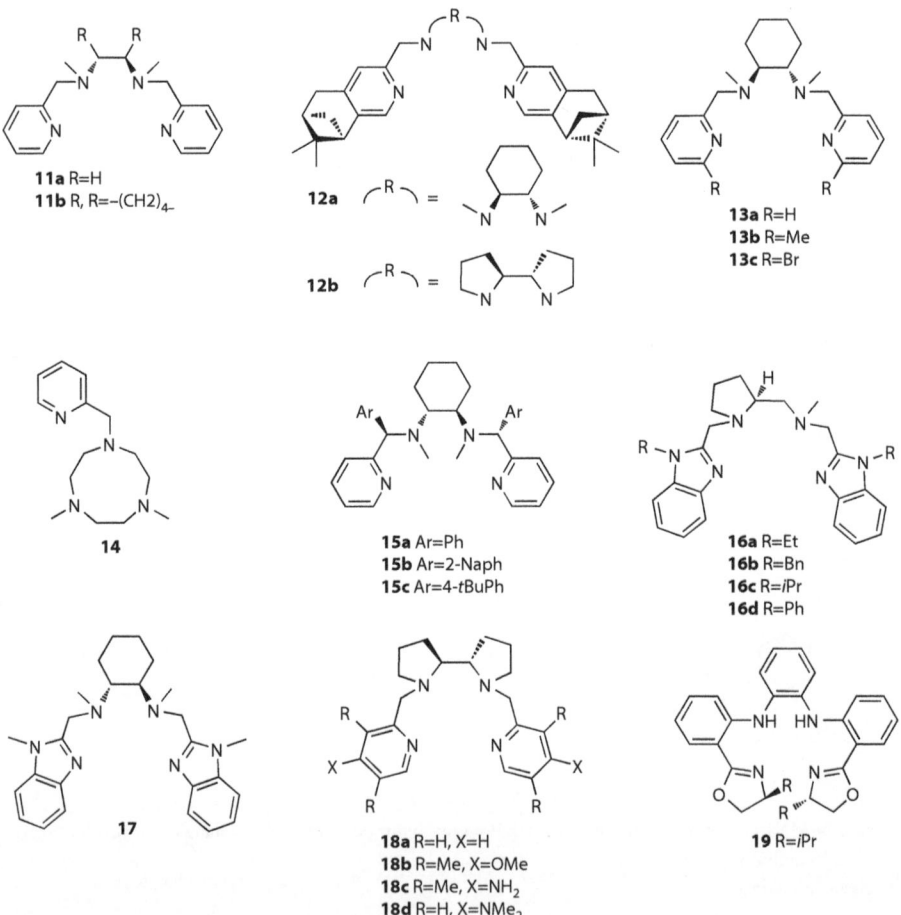

6a R=Me
6b R=iPr

7

8a R₁=R₂=Me
8b R₁=R₂= iPr
8c R₁=Me, R₂=iPr

9

10a X=Y=Me
10b X=OH, Y=H

FIGURE 2.2   Triazacyclononane-derived chiral ligands.

11a R=H
11b R, R=−(CH2)₄−

12a ⌐R⌐ =

12b ⌐R⌐ =

13a R=H
13b R=Me
13c R=Br

14

15a Ar=Ph
15b Ar=2-Naph
15c Ar=4-tBuPh

16a R=Et
16b R=Bn
16c R=iPr
16d R=Ph

17

18a R=H, X=H
18b R=Me, X=OMe
18c R=Me, X=NH₂
18d R=H, X=NMe₂

19 R=iPr

FIGURE 2.3   Aminopyridine ligands.

ligands of types **13a** and **14**) could operate as selective and efficient olefin epoxidation catalysts with H_2O_2, performing up to 1000 turnovers. The synthetic procedure developed by the authors required only 1.2 equivalents of hydrogen peroxide. High epoxide yields were achieved upon the use of an appropriate co-catalyst (AcOH, 14 equivalents with respect to substrate) [77]. The epoxidation enantioselectivity was not documented.

Later that year, Sun and co-workers reported the synthesis of three chiral manganese triflate complexes with aminopyridine chiral ligands of type **15** and the study of enantioselective oxidations of various alkenes with H_2O_2 over those catalysts [78]. Enantioselectivities in the range of 18 to 89% *ee* were reported; the highest *ee* values were achieved with α,β-unsaturated ketones—substituted chalcones in particular [78]. The synthetic protocol of Sun required the use of 5 equivalents of acetic acid as additive and as much as 6 equivalents of H_2O_2 (versus only 1.2 equivalents in Costas' protocol [77]), apparently due to pronounced peroxide decomposition in the presence of a relatively high load of the catalase-active manganese catalyst (1 mol%).

The same group later reported even higher enantioselectivities (up to 94% *ee*) and high yields (60 to >90%) upon the use of manganese triflate complexes with C_1-symmetric ligands of types **16a** and **16b** [79]. The authors found that the Mn catalyst load could be reduced to 0.2 mol%, which allowed the reduction of the H_2O_2 excess from 6 to 2 equiv—still too much for practical applications. The C_2-symmetric framework of ligand **17** also appeared productive. A series of chalcone derivatives was epoxidized with up to 96% *ee* using 0.5 mol% of the catalysts of the type [(R,R)-(**17**)Mn(OTf)$_2$] and [(R,R)-(**17**)Fe(OTf)$_2$] at $-20°C$ [80].

Even lower catalyst loads (0.1 mol%) were reported sufficient by Bryliakov and co-workers who compared the catalytic reactivities of manganese complexes [(R,R)-(**13a**)Mn(OTf)$_2$], [(S,S)-(**13a**)Mn(OTf)$_2$], and their [(**18a**)Mn(OTf)$_2$] analog derived from optically active (S,S)-bipyrrolidine [81]. The latter catalyst showed higher turnover numbers (TONs) up to 1000 and enantioselectivities. Several electron-deficient alkenes were epoxidized within 68 to 84% *ee* in the presence of acetic acid. As in the synthetic procedure of Costas et al., 1.2 to 1.3 equivalents of H_2O_2 was enough to ensure high epoxide yields (up to 100%). Moreover, the authors found that the replacement of the acetic acid additive with bulkier carboxylic acids resulted in higher epoxidation enantioselectivities [81,82]. The effect of bulkiness of carboxylic acid on the epoxidation enantioselectivity was examined systematically and found to increase in the following order: formic acid < acetic acid < *n*-butyric acid = *n*-valeric acid = *n*-hexanoic acid < *iso*-butyric < pivalic < 2-ethylhexanoic acid.

	Added acid	ee%
	Acetic	78
	Butyric	80
	Caproic	80
	Isobutyric	82
	Pivalic	86
	2-Et-hexanoic	93

[(**17**)Mn(OTf)$_2$] (0.1 mol. %)
H_2O_2 (1.2-1.4 equiv.)
RCOOH (14 equiv.)
CH$_3$CN, $-30°C$

With the use of 2-ethylhexanoic acid, two of the alkenes considered (chalcone and 2,2-dimethyl-2H-chromene-6-carbonitrile) were epoxidized with 93% *ee* over the [(**18a**)Mn(OTf)$_2$] catalyst [82]. Importantly, the epoxidation enantioselectivity could be further enhanced by the introduction of electron-donating substituents at the tetradentate ligand. With [(**18b**)Mn(OTf)$_2$] and [(**18c**)Mn(OTf)$_2$] catalysts, chalcone epoxidation occurred with 95 and 98% *ee*, respectively, [83]. Costas and co-workers synthesized pinene-derived ligands (**12a,b**) featuring the bipyrrolidine moiety and corresponding Mn triflate complexes. Manganese (II) complexes bearing a novel pinene-derived chiral aminopyridine ligand (**19** and its (*S,S*)-bipyrrolidine derived counterpart) demonstrated high efficiencies (up to 1000 turnovers) but moderate enantioselectivities in the epoxidation of various olefins (up to 73% *ee*) [84].

More recently, Gao and co-workers reported a "porphyrin-inspired" system exploiting an in situ–generated catalyst (from Mn(OTf)$_2$ and ligand **19**) [85]. In contrast to the above systems, Gao's catalyst (used at 0.2 mol%) catalyzed the epoxidation of conjugated alkenes with higher asymmetric inductions and yields than those of electron-deficient α,β-unsaturated enones; enantioselectivities up to 99% *ee* were reported [85].

The nature of active species responsible for enantioselective oxygen transfer in manganese systems, was probed [76,82,83]. On the basis of a combined EPR and enantioselectivity studies of aminopyridine Mn and Fe systems, Bryliakov and Talsi and co-workers concluded that similar intermediates should be operative in both catalyst systems [82]. The hydroperoxometal (III) species formed initially were converted to a reactive oxometal (V) intermediate; the presence of carboxylic acid facilitated this process by promoting O-O bond heterolysis [82,83].

M = Fe, Mn
L = tetradentate aminopyridine ligand

Another important role of the carboxylic acid additive is that the bulky carboxylate is present in the structure of the oxometal (V) active species and thus affects oxygen transfer enantioselectivity (see above and [81–83]). The nature of active species of hydrocarbon oxidations in the presence of Mn and Fe aminopyridine complexes is discussed in detail in Chapter 7.

Some epoxidation results obtained on manganese- and aminopyridine-based systems are listed in Table 2.3. Advantages of such systems are high efficiency, allowing very low catalyst loads (0.1 mol% or lower [79,83]), good oxidant economy, high reaction rates, possibilities of fine tuning the epoxidation enantioselectivities by rational ligand design or adjusting the bulkiness of the carboxylic acid additive. Their major limitation is narrow substrate scope, mostly restricted to electron-deficient conjugated (α,β)-enones (chalcones in particular). Other substrates are oxidized with lower enantioselectivities; for non-conjugated terminal olefins, these catalysts seem to be unsuitable.

References in the literature to other types of manganese catalysts are rather rare. A few systems using chiral Mn–porphyrin complexes operating in CH$_2$Cl$_2$ [86,87]

TABLE 2.3

Asymmetric Epoxidations with H_2O_2 Catalyzed by Chiral Aminopyridine Manganese Complexes

N	Substrate	Catalyst	Additive	Epoxide Yield (%)	ee (%)	Ref.
1	(styrene)	[(15c)Mn(OTf)$_2$]	AcOH	85	43	[78]
		[(12b)Mn(OTf)$_2$]	AcOH	95	46	[84]
		[(18a)Mn(OTf)$_2$]	EHA	92	56	[82]
		[(18b)Mn(OTf)$_2$]^a	EHA	75	61	[83]
2	(NC-cyclohexene)	[(18a)Mn(OTf)$_2$]	AcOH	93	54	[81]
3	(cis-β-methylstyrene)	6a/Mn(OAc)$_2$	—	87^b	55^b	[68]
		[(12b)Mn(OTf)$_2$]	AcOH	88	66	[84]
		[(18a)Mn(OTf)$_2$]	EHA	57	72	[82]
4	(NC-dimethylchromene)	[(12b)Mn(OTf)$_2$]	AcOH	85	66	[84]
		[(16a)Mn(OTf)$_2$]	AcOH	80	79	[79]
		[(18a)Mn(OTf)$_2$]	IBA	51	84	[81]
		[(18a)Mn(OTf)$_2$]	EHA	100	93	[82]
		[(18b)Mn(OTf)$_2$]	EHA	73	96	[83]
5	(cyclohexenone)	[(18a)Mn(OTf)$_2$]	AcOH	61	55	[82]
		[(12b)Mn(OTf)$_2$]	AcOH	65	60	[84]
6	(chalcone)	[(12b)Mn(OTf)$_2$]	AcOH	93	72	[84]
		[(15c)Mn(OTf)$_2$]	AcOH	91	78	[78]
		[(16a)Mn(OTf)$_2$]	AcOH	93	92	[79]
		[(18a)Mn(OTf)$_2$]	EHA	97	93	[82]
		[(18b)Mn(OTf)$_2$]	EHA	100	95	[83]
		[(18b)Mn(OTf)$_2$]	EHA	45	98	[83]
		[(18c)Mn(OTf)$_2$]	EHA	99	98	[83]
7	(methyl cinnamate)	[(18c)Mn(OTf)$_2$]	EHA	91	89	[83]
8	(isopropyl cinnamate)	[(16a)Mn(OTf)$_2$]	AcOH	93	74	[79]

^{a} Catalyst load: 0.01 mol%.

^{b} Complete conversion observed; a 7:1 *trans*-epoxide–*cis*-epoxide mixture was formed; *ee* for *trans*-isomer shown.

Note: EHA = 2-ethylhexanoic acid. IBA = isobutyric acid.

or in water/methanol solutions [88] were reported. In the presence of nitrogen bases, several alkenes were epoxidized with 20 to 68% *ee* with H_2O_2.

IRON AND RUTHENIUM SYSTEMS

Iron is among the most abundant metals on Earth; it is a component of many heme and non-heme enzymes and catalyzes various oxidative transformations [89,90]. For a long time, researchers pursued the goal of modeling the functions of natural enzymes with synthetic iron-containing complexes (the so-called biomimetic approach) [90]. The history of iron-based biomimetic catalysts commenced with iron–porphyrin complexes.

The first example dates back to 1979 when Groves and co-workers treated an iron (III)–porphyrin complex with iodosylbenzene to generate an active oxoiron (IV) complex capable of epoxidizing alkenes and oxidizing alkane C–H groups [91]. Few years later, the authors developed a stereoselective catalyst system based on the first chiral iron porphyrin that catalyzed the epoxidation of *p*-chlorostyrene (in 51% *ee*) with iodosylbenzene [92].

Over more than 30 years of their history, various examples of iron- and porphyrin-catalyzed oxidations with H_2O_2 and O_2 were reported [93,94]. However, for enantioselective epoxidations [17], the use of environmentally benign hydrogen peroxide and dioxygen was attained only recently [95]. Ruthenium, an analog to iron, also found various applications in asymmetric oxidation catalyst systems [97], including those using hydrogen peroxide and dioxygen. This chapter will review iron- and ruthenium-based catalyst systems.

IRON SYSTEMS

No examples [97] of iron-based catalyst systems capable of catalyzing enantioselective oxidation reactions with hydrogen peroxide were reported until 2001. Que with co-workers [98] and Jacobsen with co-workers [99] reported structurally similar iron (II) complexes that catalyzed *cis*-dihydroxylation of alkenes (with 3 to 82% *ee*) and epoxidation of alkenes (non-enantioselective) with H_2O_2 in the presence of an acetic acid additive. Both catalyst systems used aminopyridine ligands of the types **11a** and **13b**.

The first truly enantioselective aminopyridine iron-catalyzed epoxidation of olefins was proposed by Kwong and co-workers who reported an active catalyst (presumably [$Fe_2O(\mathbf{20})Cl_4$]) prepared from $FeCl_2$ and chiral ligand **20** (Figure 2.4) [100]. In the presence of acetic acid, various alkenes were epoxidized with H_2O_2 (1.5 equivalents) in 50 to 100% yield and 15 to 43% *ee* in the presence of 2.0 mol% of [$Fe_2O(\mathbf{20})Cl_4$].

In 2011, Sun with co-workers reported the catalytic properties of aminopyridine iron (II) triflate complexes with ligands of types **13a**, **15a**, and **15c** (Figure 2.3). The complexes catalyzed the epoxidation of α,β-enones (substituted chalcones) with moderate to good yields (33 to 90%) and 54 to 87% *ee* [102]. The authors' protocol implied a 2 mol% catalyst load, 5 equivalents (with respect to substrate) of acetic acid, and injection of 50% H_2O_2 (2 equivalents) via a syringe pump over 60 min [101].

More recently, Bryliakov and Talsi et al. reported enantioselective epoxidations of various alkenes with commercial 30% aqueous H_2O_2 over an iron (II) complex

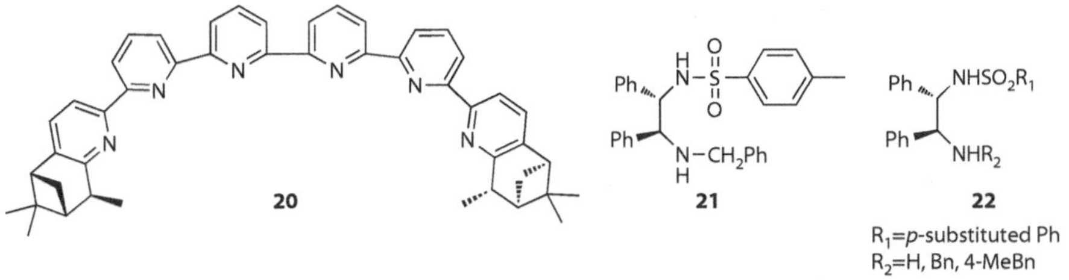

FIGURE 2.4 Chiral ligands used in iron-catalyzed enantioselective epoxidations.

[(**18a**)Fe(OTf)$_2$] with (*S,S*)–bipyrrolidine-derived ligand **18a** (Figure 2.3) [82]. The latter at 1 mol% load catalyzed the epoxidation of chalcone with H$_2$O$_2$ (2 equivalents) in 13% yield and 61% *ee*. The epoxide yield and *ee* could be increased substantially by the addition of carboxylic acid (1.1 equivalents to substrate). The enantioselectivities rose with increasing steric bulk of the acid as noted with a similar Mn-based system [82].

With the bulkiest additive, 2-ethylhexanoic acid, chalcone epoxide formed in 98% yield with 86% *ee*. Also in 2012, Sun and co-workers synthesized a series of C_1-symmetric iron (II) triflate complexes with ligands **16a**, **16c**, **16d**; catalyst [(**16a**) Fe(OTf)$_2$] demonstrated high enantioselectivities in the epoxidation of chalcone (up to 92% *ee*) and even higher for halogen-substituted chalcone derivatives and more sterically demanding α,β-enones [102].

We must note that aminopyridine iron (II) catalysts reported to date have proven inferior to their manganese counterparts (see above) in both enantioselectivity and efficiency. The only notable exception is [(**18d**)Fe(OTf)$_2$], which catalyzed the epoxidation of chalcone in the presence of 2-ethylhexanoic acid with 98% *ee*, albeit with much lower productivity (TON = 50) [103].

Some examples of enantioselective epoxidations with such systems are listed in Table 2.4. Apparently, the aminopyridine iron (V) oxo complexes in these systems are similar to those in aminopyridine manganese systems (pages 13–14) and are responsible for the enantioselective oxygen transfer [83]. In recent publications, such oxoiron (V) species were independently detected by cryospray mass spectrometry and by EPR [104,105]. The nature of active species of hydrocarbon oxidations in the presence of Mn and Fe aminopyridine complexes is discussed in more detail in Chapter 7.

A notable success in the arena of iron-catalyzed epoxidations with H$_2$O$_2$ is attributed to Beller's group. Unlike most of the aforementioned catalyst systems that rely on isolated iron complexes, Beller et al. preferred to prepare the catalytically active centers in situ by combining FeCl$_3$, dipicolynic acid (H$_2$dipic), and an organic base. After having started with achiral organic bases (such as benzylamine, 4-methylimidazole, and pyrrolidine) [106], the authors later extended their approach to asymmetric epoxidations. A number of chiral ligands (including commercially available ligands) featuring chiral 1,2-diamine and sulfonyl moieties were screened, and compound **21** emerged as an efficient chiral auxiliary for iron catalyzed enantioselective epoxidations of conjugated trans-alkenes, especially *trans*-stilbene derivatives, to form the corresponding epoxides with 40 to 94% yield and 10 to 97% *ee* [107].

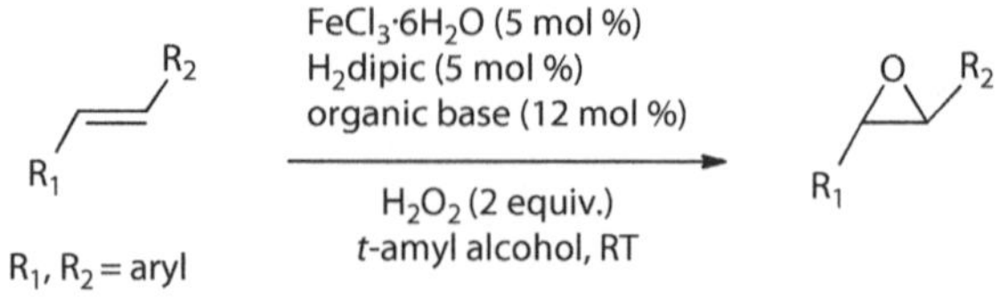

In a subsequent publication, the authors screened a series of similar chiral ligands of the common structure **22** (Figure 2.4). Based on kinetic and isotopic labeling studies, formation of a high-valence active oxo-iron intermediate was suspected; the authors detected no [18]O label incorporation into the epoxide in the course of epoxidation in the presence of an excess of $H_2{}^{18}O$, supposedly due to high reactivity of the oxoferryl intermediate [108].

TABLE 2.4

Asymmetric Epoxidations with H_2O_2 Catalyzed by Chiral Aminopyridine and Polypyridine Iron Complexes

N	Substrate	Catalyst	Additive	Epoxide Yield (%)	ee (%)	Ref.
1	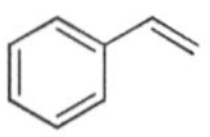	[(**18a**)Fe(OTf)$_2$]	AcOH	26	16	[82]
		[Fe$_2$O(**20**)Cl$_4$]	AcOH	95	43	[100]
2	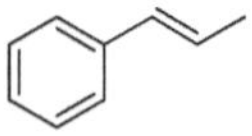	[Fe$_2$O(**20**)Cl$_4$]	AcOH	90	37	[100]
		FeCl$_2$/H$_2$dipic/**21**	—	94	28	[107]
3	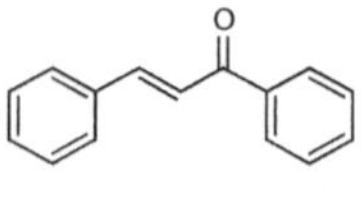	[(**18a**)Fe(OTf)$_2$]	AcOH	92	71	[82]
		[(**15c**)Fe(OTf)$_2$]	AcOH	52	78	[101]
		[(**18a**)Fe(OTf)$_2$]	EHA	98	86	[82]
		[(**16a**)Fe(OTf)$_2$]	AcOH	89	92	[102]
		[(**18d**)Fe(OTf)$_2$]	EHA	99	98	[103]
4	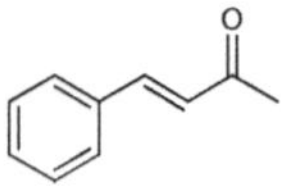	[(**18a**)Fe(OTf)$_2$]	AcOH	51	62	[82]
5	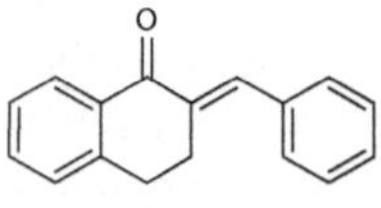	[(**16c**)Fe(OTf)$_2$]	AcOH	92	94	[102]
		[(**18d**)Fe(OTf)$_2$]	EHA	97	97	[103]
6	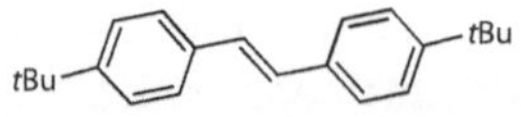	FeCl$_2$/H$_2$dipic/**21**	—	82	81	[107]
7	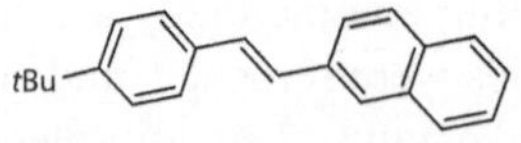	FeCl$_2$/H$_2$dipic/**21**	—	40	97	[107]

Note: EHA = 2-ethylhexanoic acid.

RUTHENIUM SYSTEMS

The first ruthenium-based enantioselective catalyst system relying on molecular oxygen as the oxygen source was reported in 1997, when Kureshy and co-workers synthesized three chiral ruthenium(III)-Schiff base complexes of type **23** (Figure 2.5), they were found to catalyze the enantioselective epoxidation of *p*-substituted styrenes (with 12 to 30% *ee*) by molecular oxygen in the presence of 1 equivalent of isobutyraldehyde as a sacrificial reductant (Table 2.5) [109]. When pyridine-*N*-oxide was added, the resulting conversions and enantioselectivities improved slightly; catalyst load of 0.3 mol% was sufficient for achieving high olefin conversions (48 to 92%) [109].

One year later, Che with co-workers synthesized a chiral ruthenium(VI) porphyrin complex **24** that catalyzed the epoxidation of several conjugated alkenes with

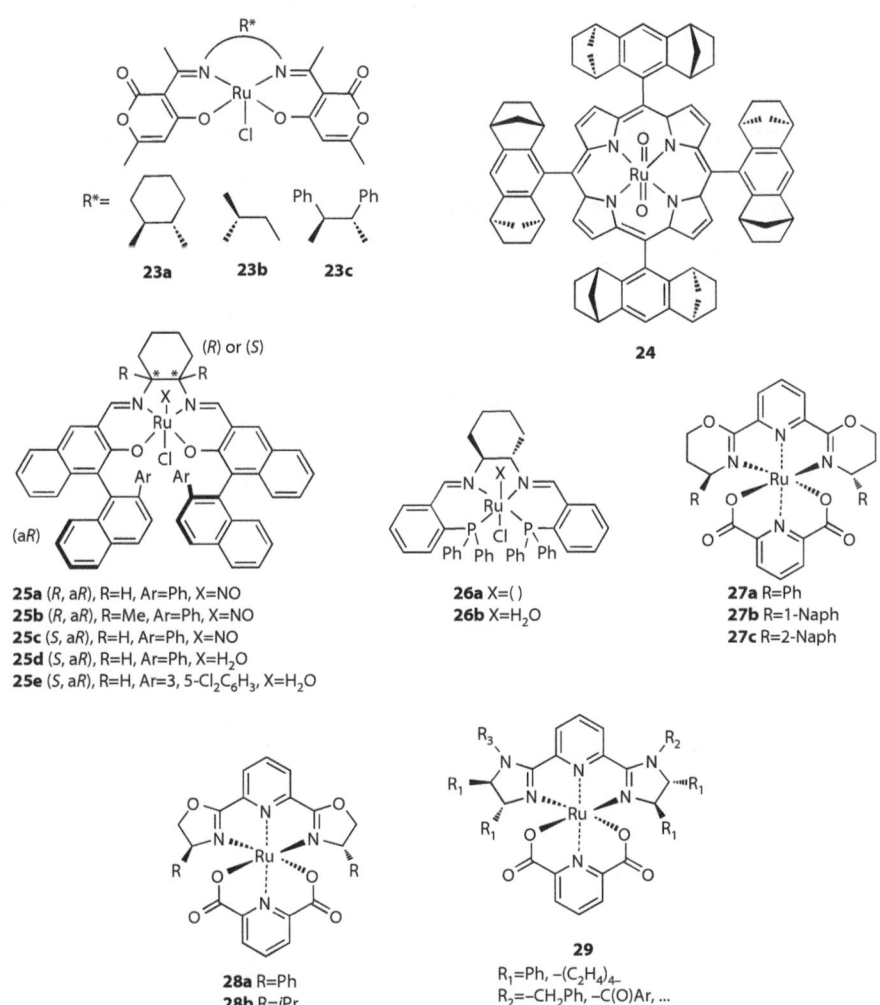

FIGURE 2.5 Chiral ruthenium catalysts.

molecular oxygen (at 8 atm pressure) without a sacrificial co-reductant [110]. Catalyst **24** demonstrated moderate enantioselectivities (52 to 73% *ee*), albeit with very low catalyst efficiency (1 to 14 turnovers, and up to 21 in one case).

In 2010, Katsuki and co-workers proposed three chiral ruthenium-salen complexes **25a-c**, featuring additional elements of axial chirality at the 3,3′ positions of the salicylidene rings (Figure 2.5) [111]. Under oxygen atmosphere (1 atm) and halogen lamp irradiation, catalyst **25c** demonstrated good to high enantioselectivities

TABLE 2.5

Asymmetric Epoxidations Catalyzed by Chiral Ruthenium Catalysts

N	Substrate	Catalyst	Oxidant	Additive	Epoxide Yield (%)	ee (%)	Ref.
1		23b	O_2	Py-*N*-O	85[a]	24	[109]
		24	O_2	—	10 TON[c]	70	[110]
		26a	H_2O_2	—	12	40	[113]
		27c	H_2O_2	—	59	48	[116]
2		25c	O_2	—	46	89	[111]
		25e	O_2	—[b]	77	93	[112]
3		23b	O_2	Py-*N*-O	55[a]	30	[109]
4		24	O_2	—	21 TON[c]	73	[110]
		25c	O_2	—	34	90	[111]
		26a	H_2O_2	—	16	25	[113]
5		26a	H_2O_2	—	55	41	[113]
6		27c	H_2O_2	—	91	84	[116]
7		27c	H_2O_2	–	>99	79	[118]

Note: Py-*N*-O = pyridine-*N*-oxide.
[a] Olefin conversion.
[b] Brine added.
[c] Turnover number reported.

(77 to 92% *ee*) for the epoxidation of both *Z* and *E* homologues of styrene, albeit with low efficiencies (5 to 15 turnovers). Using $^{18}O_2$ as the oxygen source, the authors found that the major part of oxygen in the epoxide was the ^{18}O isotope and thus confirmed that molecular oxygen is the actual terminal oxidant. The crucial role of traces of water acting as proton mediators and enhancing both the reaction rate and enantioselectivity was established [111]. The major disadvantage of that catalyst system was its low reactivity: at 5 mol% catalyst load, the epoxide yields did not exceed 26 to 75% within 48 hr.

More recently, the authors found that complexes **25d** and **25e**, each containing a coordinated molecule of water, are capable of conducting aerobic oxidation of conjugated olefins without co-reductants or light irradiation [112]. Complex **25e** with a robust salen ligand demonstrated better epoxide yields (25 to >99%) compared to **25d**, without loss in enantioselectivity (up to 94% *ee*).

The first example of ruthenium-catalyzed asymmetric epoxidations with H_2O_2 appeared in 1999 when Mezzetti and co-workers reported ruthenium(II) complexes of type **26** (Figure 2.5) with phosphonoimino and phosphonoamino *P,N,N,P*-donor chiral ligands [113–115]. The complexes used at 1 mol% load catalyzed the epoxidation of styrene, β-methylstyrenes, and 1,2-dihydronaphthalene with H_2O_2 (7 equivalents) with low-to-moderate enantioselectivities (4 to 41% *ee*). Systems of this type could also use O_2/heptaldehyde as a terminal oxidant, with lower conversion and enantioselectivity [112].

In 2004, Beller and co-workers screened a series of ruthenium(II) complexes of the types **27, 28** (and a few similar structures), prepared in situ by mixing a ruthenium precursor with the corresponding chiral ligand and 2,6-pyridinedicarboxylic acid [116]. The resulting catalysts (5 mol%), catalyzed the enantioselective epoxidation of styrene and its homologues and derivatives with 3 equivalents of 30% hydrogen peroxide, with low-to-good enantiomeric excesses (3 to 84% *ee*) [116]. As a rule, raising the steric bulk of the R substituents in complexes **27** and **28** resulted in an increase of epoxidation enantioselectivity [116].

In subsequent publications, the authors screened a series of structures of the chiral moiety (29 new complexes synthesized) [117], studied the effects of various solvents, and considered the reaction mechanism. Hammett analysis showed the possibility of a charge transfer complex [118]. The same group studied a series of structurally related pyridine–*bis*(imidazoline) complexes of ruthenium(II) of type **29** that were prepared and screened as catalysts in asymmetric epoxidations with H_2O_2, to feature good yields (63 to >99%) and low to moderate enantioselectivities (1 to 71% *ee*) [119–121].

To date, ruthenium-catalyzed enantioselective epoxidation with H_2O_2 or O_2 remains poorly developed. In addition to relatively low enantioselectivities and narrow substrate scope (mostly limited with styrenes), ruthenium catalysts demonstrated significantly lower reactivities compared to iron and manganese counterparts. At similar catalyst loads, concentrations, and temperature conditions, Ru-catalyzed reactions sometimes required days versus hours or minutes for Fe- and Mn-catalyzed reactions. In effect, in most cases Ru-based systems required higher catalyst loads (at least 5 mol%) to accomplish the reaction within foreseeable time. It is not surprising that the high Ru concentrations in the reaction mixtures resulted in substantial decomposition of H_2O_2, leading to three- to seven-fold overconsumption of H_2O_2.

TITANIUM SYSTEMS

Titanium is one of the least costly transition metals (and the seventh most abundant metal on Earth). The products of its hydrolysis are non-toxic—in sharp contrast to those of other transition metals such as Cr, Ni, V, Mn, etc. Its relative inertness toward redox processes and rich potential for tuning its activity and selectivity by rational ligand design make titanium a welcome protagonist for virtually any enantioselective catalytic transformation [122]. It is worth mentioning that one of the first asymmetric catalysts, the Sharpless system for the enantioselective epoxidation of allylic alcohols, exploited a catalyst formed in situ from titanium alkoxide and chiral tartrate [7].

The first titanium-based catalyst system for the enantioselective epoxidation of olefins with H_2O_2 was proposed in 2005 by Katsuki and co-workers who synthesized a bis(μ-oxo) titanium(IV)– salalen complex with the "second-generation" salalen ligand **30** (Figure 2.6) that efficiently conducted the epoxidation of olefins with 30% hydrogen peroxide [123]. The epoxidation of several conjugated alkenes occurred with high yields and good to excellent enantioselectivities (82 to >99% ee) (Table 2.6).

In the standard epoxidation procedure, 1.01 equivalents of H_2O_2 and only 1 mol% catalyst load were used; in one case, the catalyst performed 4600 turnovers with no loss of enantioselectivity. Importantly, the titanium–salalen-based system efficiently epoxidized a series of aliphatic non-activated olefins (including terminal olefins) with moderate to good yields (49 to 99%) and high enantioselectivities (70 to 97% ee) [124]. The system required no co-catalytic additives.

In 2006, the same group developed a series of synthetically more easily available chiral salan ligands of type **31** that could also serve as efficient chiral inducers in enantioselective epoxidation of olefins with H_2O_2, catalyzed by binuclear titanium–salan complexes [125]. The complexes required higher catalyst loads (5 mol%), while the yields and enantioselectivities were somewhat lower as compared to titanium–salalen complexes (Table 2.6).

The catalytic performance was improved by introducing *ortho* substituents in the aryl rings at the C3 and C3′ positions (to yield ligand **32**): this modification resulted in higher yields and enantioselectivities [126,127]. Complex **32b** demonstrated higher enantioselectivities (up to >99% ee); the use of a phosphate buffer (pH 7.4 to 8.0) improved the epoxide yield and catalytic efficiency (so that only 1 mol% of the catalyst was sufficient).

The same group found a convenient way to improve the enantioselectivity of titanium–salan-based catalyst systems in the epoxidation of styrenes. In particular, the 1,2-diaminocyclohexane chiral moiety was replaced with the *L*-proline derived diamine to yield C_1 symmetric ligands **33** [128]. The catalysts were generated in situ by combining 10 mol% of Ti(O*i*Pr)$_4$ and 10 mol% of **33** to epoxidize substituted styrenes with 97 to 98% ee [129]. In subsequent publications, titanium–salalen [126, 127] and in situ–prepared titanium–salan catalysts [129] were applied in the epoxidation of *cis*-alkenylsilanes [126] and (Z)-enol esters [130], with generally good yields (85 to 99%) and good to high (63 to >99% ee) enantioselectivities.

Sun with co-workers developed a synthetic approach to binaphthol-derived salalen and salan ligands of types **34** and **35** [131]. The resulting catalysts (used at 10 mol% load, formed in situ from Ti(O*i*Pr)$_4$ and ligands **34** or **35**) catalyzed the epoxidation

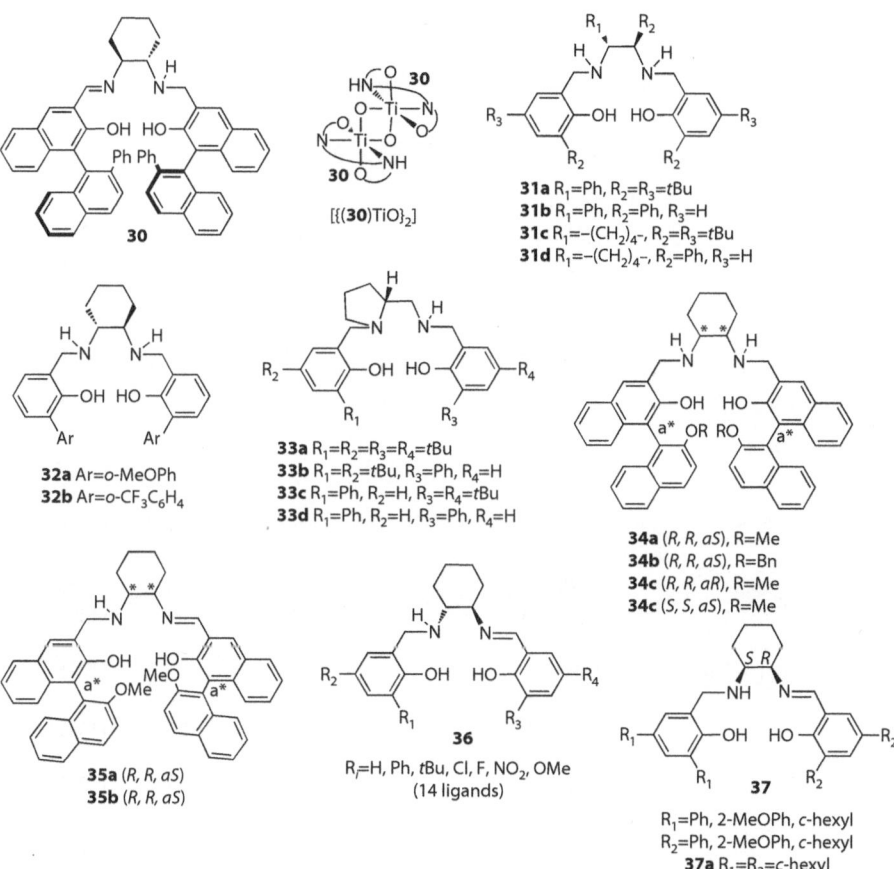

FIGURE 2.6 Chiral ligands used in titanium-catalyzed enantioselective epoxidations.

of several conjugated olefins with good to high optical yields (generally 44 to 82% *ee* and 99% *ee* in one case) with 50% H_2O_2 (3 equivalents).

Berkessel and co-workers developed a synthetic procedure for the preparation of non-symmetrical salalen ligands of type **36** [132]. The titanium catalysts derived therefrom demonstrated acceptable yields with electron-rich olefins; for non-conjugated alkenes, the yields were below 10%. Later, Sun's group reported the synthesis of biaryl-bridged salalen ligands and an intramolecular dinuclear titanium complex prepared therefrom. The latter was a highly efficient catalyst of olefin epoxidation [133]. In turn, Berkessel and co-workers prepared a series of salalen complexes with ligands of type **37** starting from optically pure *cis*-1,2-diaminocyclohexanes that provided high yields and enantioselectivities for non-conjugated terminal olefins (up to 95% *ee* for 1-octene with catalyst **37a**) [134].

Much debate has surrounded the natures of the true catalytically active sites and oxygen-transferring species operating in the titanium–salan- and titanium–salalen-based catalyst systems. Katsuki and co-workers isolated and characterized (via x-ray) a binuclear μ-oxo-μ-peroxo titanium–salan complex formed in the reaction of

TABLE 2.6

Titanium-Catalyzed Asymmetric Epoxidations with H_2O_2

N	Substrate	Catalyst	Additive	Epoxide Yield (%)	ee (%)	Ref.
1		[{(30)TiO}$_2$]	—	90	93	[123]
		[{(31d)TiO}$_2$]	—	47	82	[125]
		[{(33d)TiO}$_2$]	—	71	98	[128]
		[{(34a)TiO}$_2$]	—	78	80	[131]
2		[{(33d)TiO}$_2$]	—	80	98	[128]
3		[{(35b)TiO}$_2$]	—	95	80	[131]
4		[{(30)TiO}$_2$]	—	>99	>99	[123]
		[{(31d)TiO}$_2$]	—	79	98	[125]
5	Ph	[{(30)TiO}$_2$]	—	64	88	[123]
		[{(31d)TiO}$_2$]	—	69	90	[125]
6	n-C$_6$H$_{13}$	[{(30)TiO}$_2$]	—	85	82	[124]
		[{(31d)TiO}$_2$]	—	25	55	[125]
		[{(37a)TiO}$_2$]	—	82	95	[134]
7	n-C$_6$H$_{13}$	[{(30)TiO}$_2$]	—	85	74	[124]
8	SiMe$_3$	[{(30)TiO}$_2$]	—	87	>99	[129]

the starting (also binuclear) catalyst with H_2O_2. The authors believed the μ-oxo-μ-peroxo dimer to be a precursor of the elusive active epoxidizing species [135].

Using high resolution mass spectrometry, Berkessel and co-workers identified the titanium–salalen complexes formed in situ from the salalen ligand and Ti(OiPr)$_4$ as a mononuclear species [132]. On the basis of kinetic and mass spectrometric studies, they predicted that the active epoxidizing species could also be mononuclear peroxotitanium(IV)–salalen species [136].

It is worth mentioning that the presence of at least one N-H moiety seems essential for efficient catalysis. Neither Ti(salen) [136] nor the *N,N*′-dimethylated derivative of **27** [126] displayed catalytic activity in the epoxidation of olefins under comparable conditions. Titanium–salan complexes were studied as catalysts of asymmetric oxidation of sulfides (see Chapter 3). Apparently structurally similar intermediates should operate in both processes. The discussion of the nature of active oxidizing species will be continued in Chapter 3.

SYSTEMS BASED ON OTHER METALS

Besides Mn, Fe, Ru, Ti, a few other catalyst systems are able to conduct asymmetric epoxidation of olefins with hydrogen peroxide. Since 1987, Strukul and co-workers have developed asymmetric versions of their diphosphine-modified platinum(II) based systems [137]. Using the chiral ligands of type **38** (Figure 2.7), the authors synthesized complexes having the general formula (**37**) [Pt(CF$_3$)(CH$_2$Cl$_2$)][BF$_4$] that, at <1 mol% loads, catalyzed the epoxidation of propene and 1-octene with low to good yields and 31 to 41% ee, performing up to 108 turnovers within 72 hr [137]. Later, a series of new chiral diphosphines were developed. Complex [(**39**)Pt(CF$_3$)(CH$_2$Cl$_2$)][BF$_4$] was shown to catalyze the epoxidation of α,β-unsaturated ketones with low yields and ees. The enantioselectivity decreased over the course of the catalytic reaction [138].

In 2006, Strukul and co-workers studied the epoxidation of terminal olefins with H$_2$O$_2$ (1 equivalent) catalyzed by a series of platinum catalysts **40** (2 mol%) [139]. The catalysts demonstrated moderate to good yields (27 to 98%) and good to high enantioselectivities (58 to 87% ee and 98% ee in one case) [139]. Ligands **38a** and **41** ensured the highest enantioselectivities. The latter increased at reduced temperatures (to −10°C); the reactions were slow and required 20 to 48 hr to complete.

The selectivity of the catalyst system toward the epoxidation of terminal double bond was remarkable: 2-methyl-1,4-pentadiene yielded exclusively the 4,5-epoxide; a conjugated olefin (styrene) was not epoxidized at all. This is a rare example

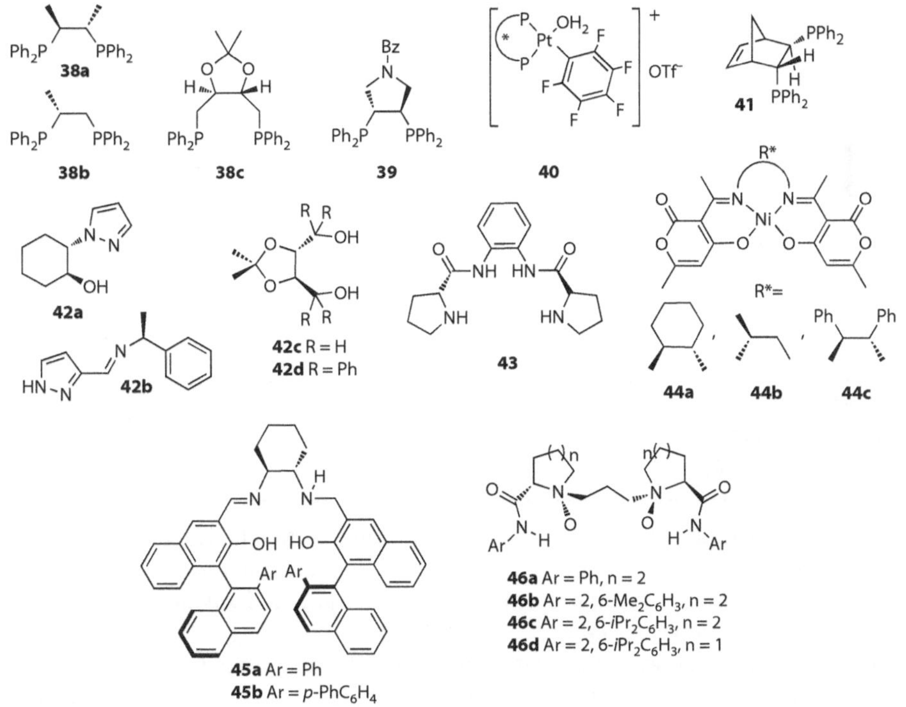

FIGURE 2.7 Chiral ligands and complexes of other metals.

TABLE 2.7

Asymmetric Epoxidations with H$_2$O$_2$ Catalyzed by Various Metals

N	Substrate	Catalyst	Additive	Epoxide Yield (%)	ee (%)	Ref.
1		[(**38a**)Pt(CF$_3$)(CH$_2$Cl$_2$)][BF$_4$]	—	Not reported	41	[137]
2		MTO–(*R*)-(+)-1-Phenyl ethylamine	—	8	20	[142]
3		MTO–(*R*)-(+)-1-Phenyl ethylamine	—	7.7	35	[142]
4		[(**38a**)Pt(OH$_3$)(C$_6$F$_5$)][CF$_3$SO$_3$]	—	79	75	[139]
5	n-C$_6$H$_{13}$	[(**38a**)Pt(CF$_3$)(CH$_2$Cl$_2$)][BF$_4$]	—	Not reported	36	[137]
		[(**38a**)Pt(OH$_3$)(C$_6$F$_5$)][CF$_3$SO$_3$]	—	88	79	[139]
6		[(**38a**)Pt(OH$_3$)(C$_6$F$_5$)][CF$_3$SO$_3$]	—	66[a]	98	[139]
7		[(**38a**)Pt(OH$_3$)(C$_6$F$_5$)][CF$_3$SO$_3$]	—[b]	56	58	[141]
		[(**38a**)Pt(OH$_3$)(C$_6$F$_5$)][CF$_3$SO$_3$]	Triton-X100[c]	61	82	[141]
8		(**43**)Cu/SiO$_2$	TEA	88[d]	83	[145]
		46c/Sc(OTf)$_3$	—	99	98	[149]
9	tBu	**46c**/Sc(OTf)$_3$	—	99	96	[149]

Note: MTO = methyltrioxorhenium. TEA = triethylamine.

[a] Selectivity toward terminal epoxide: 100%.

[b] In dichloroethane.

[c] In water–surfactant media.

[d] Conversion reported.

of a nucleophilic metal-based oxidant in transition metal-catalyzed epoxidations [23,139,140]. This reaction was also realized in water-surfactant media that in some cases allowed a significant improvement in the asymmetric induction compared to the use of organic solvents [141].

Corma's group reported the first enantioselective epoxidation of alkenes with H$_2$O$_2$ on rhenium complexes [142]. The active catalysts were generated in situ from methyltrioxorhenium (MTO) and a series of chiral amines (10 mol% each); the epoxidation of 1-methylcyclohexene, β-methylstyrenes, and α-pinene proceeded with 4

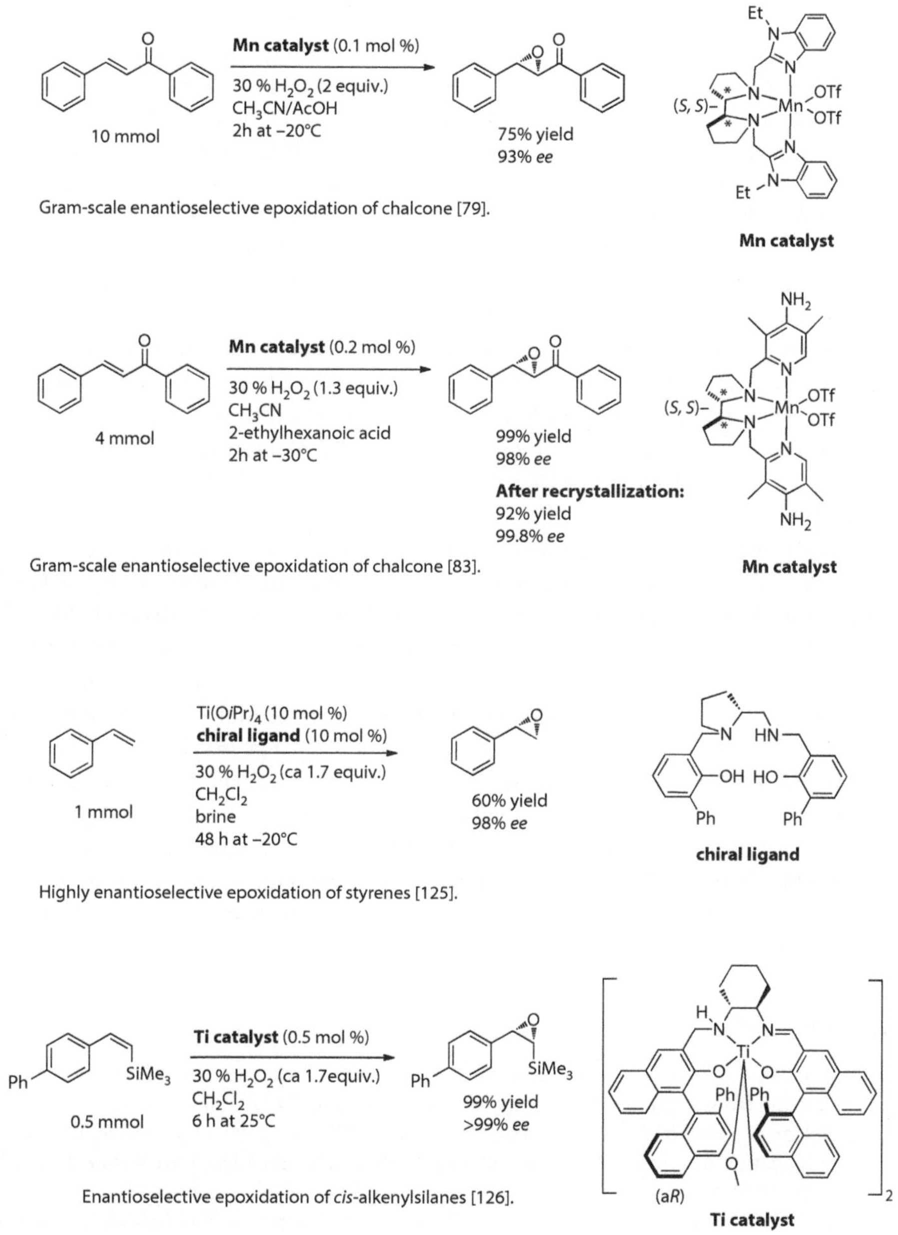

Gram-scale enantioselective epoxidation of chalcone [79].

Gram-scale enantioselective epoxidation of chalcone [83].

Highly enantioselective epoxidation of styrenes [125].

Enantioselective epoxidation of *cis*-alkenylsilanes [126].

FIGURE 2.8 Examples of preparative asymmetric epoxidations.

to 36% *ee*. However, the conversions were low (9 to 59% within 7 hr), and significant amounts of diol formed.

Herrmann and co-workers tested a number of chiral pyrazoles and diols of type **42** in the MTO-catalyzed epoxidations of Z-β-methylstyrene with H$_2$O$_2$. Despite the high (12 mol%) catalyst load, the conversions and optical yields were generally low

(25 to 30%) [143]. A series of chiral pyridine–ester and pyridine–amide ligands were tested in the MTO-catalyzed epoxidations with urea hydroperoxide; the conversions were low and the enantioselectivities were very poor (not exceeding 12%) [144].

Park and co-workers prepared heterogenized copper(II) catalysts immobilized on mesoporous silica with chiral proline-derived ligand **43** [145]. The resulting material catalyzed the epoxidation of several α,β-unsaturated ketones with 2 equivalents of H_2O_2 or urea hydroperoxide under solvent-free conditions in the presence of triethylamine additive. The reported conversions varied from 54 to 92%, and the enantioselectivities were (for chalcone epoxidation) up to 84% *ee*. The solid catalyst was recycled and reused five times without a decrease of enantioselectivity [145].

Kureshy and co-workers tested several chiral nickel(II) Schiff base complexes **44** in the epoxidation of several olefins with O_2 and isobutyraldehyde [146]. The catalysts demonstrated remarkable efficiency (over 300 turnovers), but the enantioselectivities were only moderate (14 to 41% *ee*). Very small (2 to 4% *ee*) enantioselectivities were observed in Co(salen)-catalyzed oxidation of alkenes by O_2 and isobutyraldehyde [147].

Katsuki's group contributed the first niobium-catalyzed asymmetric epoxidation of allylic alcohols with H_2O_2 and urea hydroperoxide [149]. The catalysts were generated in situ from Nb(O*i*Pr)$_5$ and chiral salan ligands of types **31c** and **45**. Moderate to good yields (40 to 82%) and *ee*s (36 to 95%) were reported using 4 mol% of the catalyst [148].

Feng and co-workers studied the epoxidation of α,β-enones (mostly substituted chalcones) with H_2O_2 (3 equivalents) in the presence of 10 mol% of scandium catalyst prepared in situ from Sc(OTf)$_3$ and chiral ligands of type **46** [149]. Various α,β-enones epoxidized with good yields (70 to 99%) and excellent enantioselectivities (83 to 99% *ee*); ligand **46c** showed the highest enantioselection.

Overall, asymmetric epoxidations with H_2O_2 or O_2 by catalyst systems based on metals other than Fe, Mn, Ti, Ru are rather rare. At the moment, none has developed sufficiently to transition from laboratory to practice. Nevertheless, it is gratifying to see that progress in this area continues and the number of metals is rapidly expanding.

REFERENCES

1. Kolb, H. C., Finn, M. G., and Sharpless, K. B. 2001. Click chemistry: diverse chemical function from a few good reactions. *Angew. Chem. Int. Ed.* 40: 2004–2021.
2. Helder, R., Hummelen, J. C., Laane, R. W. et al. 1976. Catalytic asymmetric induction in oxidation reactions: the synthesis of optically active epoxides. *Tetrahedron Lett.* 17: 1831–1834.
3. Yamada, S., Mashiko, T., and Terashima, S. 1977. (Acetylacetonato)(-)-N-alkylephedrinato-dioxomolibdenum: a new class of chiral chelate complexes which catalyze asymmetric epoxidation of allylic alcohol. *J. Am. Chem. Soc.* 99: 1988–1990.
4. Michaelson, R. C., Palermo, R. E., and Sharpless, K. B. 1977. Chiral hydroxamic acids as ligands in the vanadium-catalyzed asymmetric epoxidation of allylic alcohols by tert-butyl hydroperoxide. *J. Am. Chem. Soc.* 99: 1990–1992.
5. Kagan, H. B., Mimoun, H., Mark, C. et al. 1979. Asymmetric epoxidation of simple olefins with an optically active molybdenum(VI) peroxo complex. *Angew. Chem. Int. Ed. Engl.* 18: 485–486.

6. Tani, K., Hanafusa, M., and Otsuka, S. 1979. Asymmetric epoxidation of hydrocarbon olefins by *tert*-butyl hydroperoxide with molybdenum(IV) catalysts in the presence of optically active diols: application to the asymmetric synthesis of (3S)-oxidosqualene. *Tetrahedron Lett.* 20: 3017–3020.

7. Katsuki, T. and Sharpless, K. B. 1980. The first practical method for asymmetric epoxidation. *J. Am. Chem. Soc.* 102: 5974–5976.

8. Irie, R., Noda, K., Ito, Y. et al. 1990. Catalytic asymmetric oxidation of unfunctionalized olefins. *Tetrahedron Lett.* 31: 7345–7348.

9. Irie, R., Noda, K., Ito, Y. et al. 1991. Enantioselective epoxidation of unfunctionalized olefins using chiral (salen)manganese(III) complexes. *Tetrahedron Lett.* 32: 1055–1058.

10. Irie, R., Noda, K., Ito, Y. et al. 1991. Catalytic asymmetric epoxidation of unfunctionalized olefins using chiral (salen) manganese(III) complexes. *Tetrahedron Asymmetry* 2: 481–494.

11. Irie, R., Ito, Y., and Katsuki, T. 1991. Donor ligand effect in asymmetric epoxidation of unfunctionalized olefins with chiral salen complexes. *Synlett.* 265–266.

12. Zhang, W., Loebach, J. L., Wilson, S. R. et al. 1990. Enantioselective epoxidation of unfunctionalized olefins catalyzed by salen–manganese complexes. *J. Am. Chem. Soc.* 112: 2801–2803.

13. Zhang, W. and Jacobsen, E. N. 1991. Asymmetric olefin epoxidation with sodium hypochlorite catalyzed by easily prepared chiral manganese(III)–salen complexes. *J. Org. Chem.* 56: 2296–2298.

14. Jacobsen, E. N., Zhang, W., Muci, A. R. et al. 1991. Highly enantioselective epoxidation catalysts derived from 1,2-diaminocyclohexane. *J. Am. Chem. Soc.* 113: 7063–7064.

15. Palucki, M., Pospisil, P. J., Zhang, W. et al. 1994. Highly enantioselective, low-temperature epoxidation of styrene. *J. Am. Chem. Soc.* 116: 9333–9334.

16. Johnson, R. A. and Sharpless, K. B. 2000. Catalytic asymmetric oxidation of allylic alcohols. In *Catalytic Asymmetric Synthesis*, 2nd ed., Ojima, I., Ed. New York: John Wiley & Sons, pp. 231–285.

17. Katsuki, T. 2000. Asymmetric oxidation of unfunctionalized olefins and related reactions. In *Catalytic Asymmetric Synthesis*, 2nd ed., Ojima, I., Ed. New York: John Wiley & Sons, pp. 287–325.

18. Bonini, C. and Righi, G. 2002. A critical outlook and comparison of enantioselective oxidation methodologies of olefins. *Tetrahedron* 58: 4981–5021.

19. Cozzi, P. G. 2004. Metal–salen Schiff base complexes in catalysis: practical aspects. *Chem. Soc. Rev.* 33: 410–421.

20. Shi, Y. 2004. Organocatalytic asymmetric epoxidation of olefins by chiral ketones. *Acc. Chem. Res.* 37: 488–496.

21. McGarrigle, E. M. and Gilheany, D. G. 2005. Chromium- and manganese–salen-promoted epoxidation of alkenes. *Chem. Rev.* 105: 1563–1602.

22. Rose, E., Andrioletti, B., Zrig, S. et al. 2005. Enantioselective epoxidation of olefins with chiral metalloporphyrin catalysts. *Chem. Rev.* 34: 573–583.

23. Xia, Q. H., Ge, H. Q., Ye, C. P. et al. 2005. Advances in homogeneous and heterogeneous catalytic asymmetric epoxidation. *Chem Rev.* 105: 1603–1662.

24. Eldik, R. and Reedijk, J., Eds. 2006. *Homogeneous Biomimetic Oxidation Catalysis*, Vol. 58. Advances in Inorganic Chemistry Series. Amsterdam: Elsevier.

25. Wong, O. A. and Shi, Y. 2008. Organocatalytic oxidation: asymmetric epoxidation of olefins catalyzed by chiral ketones and iminium salts. *Chem. Rev.* 108: 3958–3987.

26. Piera, J. and Bäckvall, J. E. 2008. Catalytic oxidation of organic substrates by molecular oxygen and hydrogen peroxide by multistep electron transfer: a biomimetic approach. *Angew. Chem. Int. Ed.* 47: 3506–3523.

27. Diez, D., Nunes, M. G., Anton, A. B. et al. 2008. Asymmetric epoxidation of electron-deficient olefins. *Curr. Org. Synthesis* 5: 186–216.

28. De Faveri, G., Ilyashenko, G., and Watkinson, M. 2011. Recent advances in catalytic asymmetric epoxidation using the environmentally benign oxidant hydrogen peroxide and its derivatives. *Chem. Soc. Rev.* 40: 1722–1760.

29. Talsi, E. P. and Bryliakov, K. P. 2012. Chemo- and stereoselective C–H oxidations and epoxidations/cis-dihydroxylations with H_2O_2, catalyzed by non-heme iron and manganese complexes. *Coord. Chem. Rev.* 256: 1418–1434.

30. Russo, A., De Fusco, C., and Lattanzi, A. 2012. Organocatalytic asymmetric oxidations with hydrogen peroxide and molecular oxygen. *Chem. Cat.Chem.* 4: 901–916.

31. Yamada, T., Imagawa, K., Nagata, T. et al. 1992. Enantioselective epoxidation of unfunctionalized olefins with molecular oxygen and aldehyde catalyzed by optically active manganese(III) complexes. *Chem. Lett.* 2231–2234.

32. Yamada, T., Imagawa, K., Nagata, T. et al. 1994. Aerobic enantioselective epoxidation of unfunctionalized olefins catalyzed by optically active salen–manganese(III) complexes. *Bull. Chem. Soc. Jpn.* 67: 2248–2256.

33. Mukaiyama, T., Yamada, T., Nagata, T. et al. 1993. Asymmetric aerobic epoxidation of unfunctionalized olefins catalyzed by optically active α-alkoxycarbonyl-β-ketoiminato manganese(III) complexes. *Chem. Lett.* 327–330.

34. Nagata, T., Imagawa, K., Yamada, T. et al. 1994. Optically active β-ketoiminato manganese (III) complexes as efficient catalysts in enantioselective aerobic epoxidation of unfunctionalized olefins. *Inorg. Chim. Acta* 220: 283–287.

35. Nagata, T., Imagawa, K., Yamada, T. et al. 1994. Enantioselective aerobic epoxidation of acyclic simple olefins catalyzed by optically active β-ketoiminato–manganese(III) vomplex. *Chem. Lett.* 1259–1262.

36. Nagata, T., Imagawa, K., Yamada, T. et al. 1995. Optically active *N,N'*-bis(3-oxobutylidene)diaminato–manganese(III) complexes as novel and efficient catalysts for aerobic enantioselective epoxidation of simple olefins. *Bull. Chem. Soc. Jpn.* 68: 1455–1465.

37. Imagawa, K., Nagata, T., Yamada, T. et al. 1995. Asymmetric oxidation of sulfides with molecular oxygen catalyzed by β-oxoaldiminato–manganese(III) complexes. *Chem. Lett.* 335–336.

38. Nagata, T., Imagawa, K., Yamada, T. et al. 1995. Enantioselective aerobic oxidation of sulfides catalyzed by optically active β-oxoaldiminato–manganese(III) complexes. *Bull. Chem. Soc. Jpn.* 68: 3241–3246.

39. Rhodes, B., Rowling, S., Tidswell, P. et al. 1997. Aerobic epoxidation via alkyl-2-oxo-cyclopentane carboxylate co-oxidation with cobalt or manganese Jacobsen-type catalysts. *J. Mol. Catal. A Chem.* 116: 375–384.

40. Imagawa, K., Nagata, T., Yamada, T. et al. 1994. N-aklyl imidazoles as effective axial ligands in the aerobic asymmetric epoxidation of unfunctionalized olefins catalyzed by optically active manganese(III)–salen complex. *Chem. Lett.* 527–530.

41. Mukaiyama, T. and Yamada, T. 1995. Recent advances in aerobic oxygenation. *Bull. Chem. Soc. Jpn.* 68: 17–35.

42. Bryliakov, K. P., Kholdeeva, O. A., Vanina, M. P. et al. 2002. Role of Mn-IV species in Mn–(salen)-catalyzed enantioselective aerobic epoxidations of alkenes: an EPR study. *J. Mol. Catal. A Chem.* 178: 47–53.

43. Lee, N. H., Baik, J. S., and Han, S. B. 1997. Trapping of the dichlorocarbonyl oxide using a chiral (salen)Mn(III) complex. *Bull. Korean Chem. Soc.* 18: 796–798.

44. Guo, P. and Wong, K. Y. 1999. Enantioselective electrocatalytic epoxidation of olefins by chiral manganese Schiff-base complexes. *Electrochem. Commun.* 1: 559–563.

45. Tanaka, H., Kuroboshi, M., Takeda, H. et al. 2001. Electrochemical asymmetric epoxidation of olefins by using an optically active Mn–salen complex. *J. Electroanal. Chem.* 507: 75–81.

46. Robert, A., Tsapara, A., and Meunier, B. 1993. Brominated and chlorinated manganese chiral Schiff base complexes as epoxidation catalysts. *J. Mol. Catal.* 85: 13–19.

47. Palucki, M., Hanson, P., and Jacobsen, E. N. 1992. Asymmetric oxidation of sulfides with H_2O_2 catalyzed by (salen)Mn(III) complexes. *Tetrahedron Lett.* 33: 7111–7114.
48. Pietikäinen, P. 1994. Catalytic and asymmetric epoxidation of unfunctionalized alkenes with hydrogen peroxide and (salen)Mn(III) complexes. *Tetrahedron Lett.* 35: 941–944.
49. Irie, R., Hosoya, N., and Katsuki, T. 1994. Enantioselective epoxidation of chromene derivatives using hydrogen peroxide as a terminal oxidant. *Synlett.* 255–256.
50. Meunier, B. 1992. Metalloporphyrins as versatile catalysts for oxidation reactions and oxidative DNA cleavage. *Chem. Rev.* 92: 1411–1458.
51. de Montellano, P. R. 1987. Control of the catalytic activity of prosthetic heme by the structures of hemoproteins. *Acc. Chem. Res.* 20: 289–294.
52. Zeng, J. and Fenna, R. E. 1992. X-ray crystal structure of canine myeloperoxidase at 3Å resolution. *J. Mol. Biol.* 226: 185–207.
53. Costas, M., Mehn, M. P., Jensen, M. P. et al. 2004. Dioxygen activation at mononuclear non-heme iron-active sites: enzymes, models, and intermediates. *Chem. Rev.* 104: 939–986.
54. Steinman, A. A. 2008. Iron oxygenases: structure, mechanism and modeling. *Russ. Chem. Rev.* 77: 945–966.
55. Tanase, S. and Bowman, K. B. 2005. Selective conversion of hydrocarbons with H_2O_2 using biomimetic non-heme iron and manganese oxidation vatalysts. In *Advances in Inorganic Chemistry*, Vol. 58, Van Eldik, R. and Reedijk, J., Eds. Amsterdam: Elsevier, pp. 29–75.
56. Schwenkreis, T. and Berkessel, A. 1993. A biomimetic catalyst for the asymmetric epoxidation of unfunctionalized olefins with hydrogen peroxide. *Tetrahedron Lett.* 34: 4785–4788.
57. Berkessel, A., Frauenkron, M., Schwenkreis, T. et al. 1996. Pentacoordinated manganese(III) dihydrosalen complexes as biomimetic oxidation catalysts. *J. Mol. Catal.* 113: 321–342.
58. Shitama, H. and Katsuki, T. 2006. Asymmetric epoxidation using aqueous hydrogen peroxide as oxidant: bio-inspired construction of pentacoordinated Mn–salen complexes and their catalysis. *Tetrahedron Lett.* 47: 3203–3207.
59. Shitama, H. and Katsuki, T. 2007. Synthesis of metal–(pentadentate) salen complexes: asymmetric epoxidation with aqueous hydrogen peroxide and asymmetric cyclopropanation (salenH$_2$: *N,N'*-bis(salicylidene)ethylene-1,2-diamine). *Chem. Eur. J.* 13: 4849–4858.
60. Pietikäinen, P. 1998. Convenient asymmetric (salen)Mn(III)-catalyzed epoxidation of unfunctionalized alkenes with hydrogen peroxide using carboxylate salt co-catalysts. *Tetrahedron* 54: 4319–4326.
61. Kureshy, R. I., Khan, N. H., Abdi, S. H. R. et al. 2001. Enantioselective epoxidation of non-functionalised alkenes using a urea–hydrogen peroxide oxidant and a dimeric homochiral Mn(III)–Schiff base complex catalyst. *Tetrahedron Asymmetry* 12: 433–437.
62. Kureshy, R. I., Khan, N. H., Abdi, S. H. R. et al. 2003. Chiral Mn(III) salen complex-catalyzed enantioselective epoxidation of nonfunctionalized alkenes using urea–H_2O_2 adduct as oxidant. *J. Catal.* 219: 1–7.
63. Maity, N. C., Abdi, S. H. R., Kureshy, R. I. et al. 2011. Chiral macrocyclic salen Mn(III) complexes catalyzed enantioselective epoxidation of non-functionalized alkenes using NaOCl and urea H_2O_2 as oxidants. *J. Catal.* 277: 123–127.
64. Lane, B. S., Vogt, M., DeRose, V. J. et al. 2002. Manganese-catalyzed epoxidations of alkenes in bicarbonate solutions. *J. Am. Chem. Soc.* 124: 11946–11954.
65. Pietikäinen, P. 2001. Asymmetric Mn(III)–salen-catalyzed epoxidation of unfunctionalized alkenes with in situ-generated peroxycarboxylic acids. *J. Mol. Catal. A Chem.* 165: 73–79.

66. Garcia, M. A., Méou, A., and Brun, P. 1996. (Salen)–Mn(III)-catalyzed asymmetric epoxidation of geraniol derivatives. *Synlett.* 1049–1050.
67. Méou, A., Garcia, M. A., and Brun, P. 1999. Oxygen transfer mechanism in the Mn–salen catalysed epoxidation of olefins. *J. Mol. Catal. A Chem.* 138: 221–226.
68. Bolm, C., Kadereit, D., and Valacchi, M. 1997. Enantioselective olefin epoxidation with chiral manganese–1,4,7-triazacyclononane complexes. *Synlett.* 687–688.
69. Bolm, C., Meyer, N., Raabe, G. et al. 2000. A novel enantiopure proline-derived triazacyclononane: synthesis, structure and application of its manganese complex. *Chem. Commun.* 2435–2436.
70. Argouarch, G., Gibson, C. L., Stones, G. et al. 2002. The synthesis of chiral annulet 1,4,7-triazacyclononanes. *Tetrahedron Lett.* 43: 3795–3798.
71. Romakh, V. B., Therrien, B., Süss-Fink, G. et al. 2007. Dinuclear manganese complexes containing chiral 1,4,7-triazacyclononane-derived ligands and their catalytic potential for the oxidation of olefins, alkanes, and alcohols. *Inorg. Chem.* 46: 1315–1331.
72. Murphy, A., Dubois, G., and Stack, T. D. P. 2003. Efficient epoxidation of electron-deficient olefins with a cationic manganese complex. *J. Am. Chem. Soc.* 125: 5250–5251.
73. Murphy, A., Pace, A., and Stack, T. D. P. 2004. Ligand and pH influence on manganese-mediated peracetic acid epoxidation of terminal olefins. *Org. Lett.* 6: 3119–3122.
74. Murphy, A. and Stack, T. D. P. 2006. Discovery and optimization of rapid manganese catalysts for the epoxidation of terminal olefins. *J. Mol. Catal. A Chem.* 251: 78–88.
75. Gómez, L., Garcia-Bosch, I., Company, A. et al. 2007. Chiral manganese complexes with pinene-appended tetradentate ligands as stereoselective epoxidation catalysts. *Dalton Trans.* 5539–5545.
76. Ottenbacher, R. V., Bryliakov, K. P., and Talsi, E. P. 2010. Non-heme manganese-catalyzed asymmetric oxidation: a Lewis acid activation versus oxygen rebound mechanism. *Inorg. Chem.* 49: 8620–8628.
77. Garcia-Bosch, I., Ribas, X., and Costas, M. 2009. A broad substrate-scope method for fast, efficient and selective hydrogen peroxide epoxidation. *Adv. Synth. Catal.* 351: 348–352.
78. Wu, M., Wang, B., Wang, S. et al. 2009. Asymmetric epoxidation of olefins with chiral bioinspired manganese complexes. *Org. Lett.* 11: 3622–3625.
79. Wang, B., Miao, C., Wang, S. et al. 2012. Manganese catalysts with C_1-symmetric N_4 ligand for enantioselective epoxidation of olefins. *Chem. Eur. J.* 18: 6750–6753.
80. Wang, X., Miao, C., Wang, S. et al. 2013. Bioinspired manganese and iron complexes with tetradentate N ligands for asymmetric epoxidation of olefins. *ChemCatChem* 5: 2489–2494.
81. Ottenbacher, R. V., Bryliakov, K. P., and Talsi, E. P. 2011. Non-heme manganese complexes catalyzed asymmetric epoxidation of olefins by peracetic acid and hydrogen peroxide. *Adv. Synth. Catal.* 353: 885–889.
82. Lyakin, O. Y., Ottenbacher, R. V., Bryliakov, K. P. et al. 2012. Asymmetric epoxidations with H_2O_2 on Fe– and Mn–aminopyridine catalysts. *ACS Catal.* 2: 1196–1202.
83. Ottenbacher, R. V., Samsonenko, D. G., Talsi, E. P. et al. 2014. Highly enantioselective bioinspired olefins with H_2O_2 on aminopyridine Mn catalysts. *ACS Catal.* 4: 1599–1606.
84. Garcia-Bosch, I., Gómez, L., Polo, A. et al. 2012. Stereoselective epoxidation of alkenes with hydrogen peroxide using a bipyrrolidine-based family of manganese complexes. *Adv. Synth. Catal.* 354: 65–70.
85. Dai, W., Li, J., Li, G. et al. 2013. Asymmetric epoxidation of alkenes catalyzed by a porphyrin-inspired manganese complex. *Org. Lett.* 15: 4138–4141.
86. Vilain, S., Maillard, P., and Momenteau, M. 1994. Enantiomeric epoxidation of 4-chlorostyrene with H_2O_2 catalysed by robust chloro manganese(III) meso-5,10,1-5,20-tetrakis-2-chloro-6-(2,3,4,6-tetraacetyl-*O*-β-glucosyl)-phenyl porphyrins. *Chem. Commun.* 1697–1698.

87. Vilain-Deshayes, S., Maillard, P., and Momenteau, M. 1996. Enantiomeric epoxidation of 4-chlorostyrene with H_2O_2 catalysed by robust chloro manganese(III)-5,10,15,20-tet-rakis-2-chloro-6-(2,3,4,6-tetraacetyl-O-β-D-glucosyl)-phenyl porphyrins. *J. Mol. Catal.* 113: 201–208.

88. Srour, H., Le Maux, P., and Simonneaux, G. 2012. Enantioselective manganese–porphy-rin-catalyzed epoxidation and C–H hydroxylation with hydrogen oeroxide in water and methanol solutions. *Inorg. Chem.* 51: 5850–5856.

89. Que, Jr., L. 2007. The road to non-heme oxoferryls and beyond. *Acc. Chem. Res.* 40: 493–500.

90. Que, Jr., L. and Tolman, W. B. 2008. Biologically inspired oxidation catalysis. *Nature* 455: 333–340.

91. Groves, J. T., Nemo, T. E., and Myers, R. S. 1979. Hydroxylation and epoxidation cata-lyzed by iron–porphine complexes: oxygen transfer from iodosylbenzene. *J. Am. Chem. Soc.* 101: 1032–1033.

92. Groves, J. T. and Myers, R. S. 1983. Catalytic asymmetric epoxidations with chiral iron porphyrins. *J. Am. Chem. Soc.* 105: 5791–5796.

93. Lane, B. S. and Burgess, K. 2003. Metal-catalyzed epoxidations of alkenes with hydro-gen peroxide. *Chem. Rev.* 103: 2457–2473.

94. Punniyamurthy, T., Velusamy, S., and Iqbal, J. 2005. Recent advances in transition metal-catalyzed oxidation of organic substrates with molecular oxygen. *Chem. Rev.* 105: 2329–2363.

95. Le Maux, P., Srour, H. F., and Simonneaux, G. 2012. Enantioselective water-soluble iron–porphyrin-catalyzed epoxidation with aqueous hydrogen peroxide and hydroxyl-ation with iodobenzene diacetate. *Tetrahedron* 68: 5824–5828.

96. Chatterjee, D. 2008. Asymmetric epoxidation of unsaturated hydrocarbons catalyzed by ruthenium complexes. *Coord. Chem. Rev.* 252: 176–198.

97. Francis, M. B. and Jacobsen, E. N. 1999. Discovery of novel catalysts for alkene epoxi-dation from metal-binding combinatorial libraries. *Angew. Chem. Int. Ed.* 38: 937–941.

98. Costas, M., Tipton, A. K., Chen, K. et al. 2001. Modeling Rieske dioxygenases: first example of iron-catalyzed asymmetric *cis*-dihydroxylation of olefins. *J. Am. Chem. Soc.* 123: 6722–6723.

99. White, M. C., Doyle, A. G., and Jacobsen, E. N. 2001. A synthetically useful, self-assembling MMO mimic system for catalytic alkene epoxidation with aqueous H_2O_2. *J. Am. Chem. Soc.* 123: 7194–7195.

100. Yeung, H. L., Sham, K. C., Tsang, C. S. et al. 2008. A chiral iron–sexipyridine com-plex as a catalyst for alkene epoxidation with hydrogen peroxide. *Chem. Commun.* 3801–3803.

101. Wu, M., Miao, C., Wang, S. et al. 2011. Chiral bioinspired non-heme iron complexes for enantioselective epoxidation of α,β-unsaturated ketones. *Adv. Synth. Catal.* 353: 3014–3022.

102. Wang, B., Wang, S., Xia, C. et al. 2012. Highly enantioselective epoxidation of multi-substituted enones catalyzed by non-heme iron catalysts. *Chem. Eur. J.* 18: 7332–7335.

103. Cussó, O., Garcia-Bosch, I., Ribas, X. et al. 2013. Asymmetric epoxidation with H_2O_2 by manipulating the electronic properties of non-heme iron catalysts. *J. Am. Chem. Soc.* 135: 14871–14878.

104. Prat, I., Mathieson, J. S., Güell, M. et al. 2011. Observation of Fe(V) = O using variable temperature mass spectrometry and its enzyme-like C–H and C=C oxidation reactions. *Nat. Chem.* 3: 788–793.

105. Lyakin, O. Y., Prat, I., Bryliakov, K. P. et al. 2012. EPR detection of Fe(V)=O active species in non-heme iron-catalyzed oxidations. *Catal. Commun.* 29: 105–108.

106. Anilkumar, G., Bitterlich, B., Gelalcha, F. G. et al. 2007. An efficient biomimetic Fe-catalyzed epoxidation of olefins using hydrogen peroxide. *Chem. Commun.* 289–291.

107. Gelalcha, F. G., Bitterlich, B., Anilkumar, G. et al. 2007. Iron-catalyzed asymmetric epoxidation of aromatic alkenes using hydrogen peroxide. *Angew. Chem. Int. Ed.* 46: 7293–7296.
108. Gelalcha, F. G., Anilkumar, G., Tse, M. K. et al. 2008. Biomimetic iron-catalyzed asymmetric epoxidation of aromatic alkenes by using hydrogen peroxide. *Chem. Eur. J.* 14: 7687–7698.
109. Kureshy, R. I., Khan, N. H., Abdi, S. H. R. et al. 1997. Chiral Ru(III) metal complex-catalyzed aerobic enantioselective epoxidation of styrene derivatives with co-oxidation of aldehyde. *J. Mol. Catal. A Chem.* 124: 91–97.
110. Lai, T. S., Zgang, R., Cheung, K. K. et al. 1998. Aerobic enantioselective alkene epoxidation by a chiral trans-dioxo(D_4-porphyrinato) ruthenium(VI) complex. *Chem. Commun.* 1583–1584.
111. Tanaka, H., Nishikawa, H., Uchida, T. et al. 2010. Photopromoted Ru-catalyzed asymmetric aerobic sulfide oxidation and epoxidation using water as a proton transfer mediator. *J. Am. Chem. Soc.* 132: 12034–12041.
112. Koya S., Nishioka, Y., Mizoguchi, H. et al. 2012. Asymmetric epoxidation of conjugated olefins with dioxygen. *Angew. Chem. Int. Ed.* 51: 8243–8246.
113. Stoop, R. M. and Mezzetti, A. 1999. Asymmetric epoxidation of olefins. *Green Chem.* 39–41.
114. Stoop, R. M., Bachmann, S., Valentini, M. et al. 2000. Ruthenium(II) complexes with chiral tetradentate P_2N_2 ligands catalyze the asymmetric epoxidation of olefins with H_2O_2. *Organometallics* 19: 4117–4126.
115. Bonaccorsi, C., Mezzetti, A. 2006. Ruthenium complexes with chiral tetradentate PNNP ligands in asymmetric catalytic atom-transfer reactions. *Curr. Org. Chem.* 10: 225–240.
116. Tse, M. K., Döbler, C., Bhor, S. et al. 2004. Development of a ruthenium-catalyzed asymmetric epoxidation procedure with hydrogen peroxide as the oxidant. *Angew. Chem. Int. Ed.* 43: 5255–5260.
117. Tse, M. K., Bhor, S., Klawonn, M. et al. 2006. Ruthenium-catalyzed asymmetric epoxidation of olefins using H_2O_2. Part I: Synthesis of new chiral N,N,N-tridentate pybox and pyboxazine ligands and their ruthenium complexes. *Chem. Eur. J.* 12: 1855–1874.
118. Tse, M. K., Bhor, S., Klawonn, M. et al. 2006. Ruthenium-catalyzed asymmetric epoxidation of olefins using H_2O_2. Part II: Catalytic activities and mechanisms. *Chem. Eur. J.* 12: 1875–1888.
119. Bhor, S., Anilkumar, G., Tse, M. K. et al. 2005. Synthesis of a new Chiral *N,N,N*-tridentate pyridine bisimidazoline ligand library and its application in Ru-catalyzed asymmetric epoxidation. *Org. Lett.* 7: 3393–3396.
120. Anilkumar, G., Bhor, S., Tse, M. K. et al. 2005. Synthesis of a novel class of chiral N,N,N-tridentate pyridine bisimidazoline ligands and their application in Ru-catalyzed asymmetric epoxidations. *Tetrahedron Asymmetry* 16: 3536–3561.
121. Tse, M. K., Jiao, H., Anilkumar, C. et al. M. 2006. Synthetic, spectral, and catalytic activity studies of ruthenium bipyridine and terpyridine complexes: implications in the mechanism of the ruthenium(pyridine-2,6-bisoxazoline)(pyridine-2,6-dicarboxylate)-catalyzed asymmetric epoxidation of olefins utilizing H_2O_2. *J. Organomet. Chem.* 691: 4419–4433.
122. Ramón, D. J. and Yus, M. 2006. In the arena of enantioselective synthesis, titanium complexes wear the laurel wreath. *Chem. Rev.* 106: 2126–2208.
123. Matsumoto, K., Sawada, Y., Saito, B. et al. 2005. Construction of pseudo-heterochiral and homochiral di-μ-oxotitanium (Schiff base) dimers and enantioselective epoxidation using aqueous hydrogen peroxide. *Angew. Chem. Int. Ed.* 44: 4935–4939.
124. Sawada, Y., Matsumoto, K., and Katsuki, T. 2007. Titanium-catalyzed asymmetric epoxidation of non-activated olefins with hydrogen peroxide. *Angew. Chem. Int. Ed.* 46: 4559–4561.

125. Sawada, Y., Matsumoto, K., Kondo, S. et al. 2006. Titanium–salan-catalyzed asymmetric epoxidation with aqueous hydrogen peroxide as the oxidant. *Angew. Chem. Int. Ed.* 45: 3478–3480.

126. Matsumoto, K., Sawada, Y., and Katsuki, T. 2006. Catalytic enantioselective epoxidation of unfunctionalized olefins: utility of a Ti(O*i*-Pr)$_4$–salan–H$_2$O$_2$ system. *Synlett.* 3545–3547.

127. Shimada, Y., Kondo, S., Ohara, Y. et al. 2007. Titanium-catalyzed asymmetric epoxidation of olefins with aqueous hydrogen peroxide: remarkable effect of phosphate buffer on epoxide yield. *Synlett.* 2445–2447.

128. Matsumoto, K., Oguma, T., and Katsuki, T. 2009. Highly enantioselective epoxidation of styrenes catalyzed by proline-derived C_1–symmetric titanium(salan) complexes. *Angew. Chem. Int. Ed.* 48: 7432–7435.

129. Matsumoto, K., Kubo, T., and Katsuki, T. 2009. Highly enantioselective epoxidation of *cis*-alkenylsilanes. *Chem. Eur. J.* 15: 6573–6575.

130. Matsumoto, K., Feng, C., Handa, S. et al. 2011. Asymmetric epoxidation of (Z)-enol esters catalyzed by titanium(salalen) complex with aqueous hydrogen peroxide. *Tetrahedron* 67: 6474–6478.

131. Xiong, D., Wu, M., Wang, S. et al. 2010. Synthesis of salan (salalen) ligands derived from binaphthol for titanium-catalyzed asymmetric epoxidation of olefins with aqueous H$_2$O$_2$. *Tetrahedron Asymmetry* 21: 374–378.

132. Berkessel, A., Brandenburg, M., Leitterstorf, E. et al. 2007. A pactical and versatile access to dihydrosalen (salalen) ligands: highly enantioselective titanium in situ catalysts for asymmetric epoxidation with aqueous hydrogen peroxide. *Adv. Synth. Catal.* 349: 2385–2391.

133. Xiong, D., Hu, X., Wang, S. et al. 2011. Biaryl-bridged salalen ligands and their application in titanium-catalyzed asymmetric epoxidation of olefins with aqueous H$_2$O$_2$. *Eur. J. Org. Chem.* 4289-4292.

134. Berkessel, A., Günther, T., Wang, Q. et al. 2013. Titanium–salalen catalysts based on *cis*-1,2-diaminocyclohexane: enantioselective epoxidation of terminal non-conjugated olefins with H$_2$O$_2$. *Angew. Chem. Int. Ed.* 52: 8467–8471.

135. Kondo, S., Saruhashi, K., Seki, K. et al. 2008. A μ-oxo-μ-η^2:η^2-peroxo titanium complex as a reservoir of active species in asymmetric epoxidation using hydrogen peroxide. *Angew. Chem. Int. Ed.* 47: 10195–10198.

136. Berkessel, A., Brandenburg, M., and Schäfer, M. 2008. Mass-spectrometric and kinetic studies on the mechanism and degradation pathways of titanium–salalen catalysts for asymmetric epoxidation with aqueous hydrogen peroxide. *Adv. Synth. Catal.* 350: 1287–1294.

137. Sinigalia, R., Michelin, R. A., Pinna, F. et al. 1987. Asymmetric epoxidation of simple olefins catalyzed by chiral diphosphine-modified platinum(II) complexes. *Organometallics* 6: 728–734.

138. Baccin, C., Gusso, A., Pinna, F. et al. 1995. Platinum-catalyzed oxidations with hydrogen peroxide: the (enantioselective) epoxidation of α,β-unsaturated ketones. *Organometallics* 14: 1161–1167.

139. Colladon, M., Scarso, A., Sgarbossa, P. et al. 2006. Asymmetric epoxidation of terminal alkenes with hydrogen peroxide catalyzed by pentafluorophenyl Pt^{II} complexes *J. Am. Chem. Soc.* 128: 14006–14007.

140. Colladon, M., Scarso, A., Sgarbossa, P. et al. 2007. Regioselectivity and diasteroselectivity in Pt(II)-mediated "green" catalytic epoxidation of terminal alkenes with hydrogen peroxide: mechanistic insight into a peculiar substrate selectivity. *J. Am. Chem. Soc.* 129: 7680–7689.

141. Colladon, M., Scarso, A., and Strukul, G. 2007. Toward a greener epoxidation method: use of water-surfactant media and catalyst recycling in the platinum-catalyzed asymmetric epoxidation of terminal alkenes with hydrogen peroxide. *Adv. Synth. Catal.* 349: 797–801.

142. Sabater, M. J., Domine, M. E., and Corma, A. 2002. Highly stable chiral and achiral nitrogen-base adducts of methyltrioxorhenium(VII) as catalysts in the epoxidation of alkenes. *J. Catal.* 210: 192–197.

143. Haider, J. J., Kratzer, R. M., Herrmann, W. A. et al. 2004. On the way to chiral epoxidations with methyltrioxorhenium(VII)-derived catalysts. *J. Organomet. Chem.* 689: 3735–3740.

144. Da Palma-Carreiro, E., Yong-En, G., and Burke, A. J. 2005. Approaches toward catalytic asymmetric epoxidations with methyltrioxorhenium(VII) (MTO): synthesis and evaluation of chiral non-racemic 2-substituted pyridines. *J. Mol. Catal. A Chem.* 235: 285–292.

145. Prasetyanto, E. A., Hasan Khan, N., Seo, H. U. et al. 2010. Asymmetric epoxidation of α,β-unsaturated ketones over heterogenized chiral proline diamide complex catalyst in the solvent-free condition. *Top. Catal.* 53: 1381–1386.

146. Kureshy, R. I., Khan, N. H., Abdi, S. H. R. et al. 1998. Aerobic, enantioselective epoxidation of non-functionalized olefins catalyzed by Ni(II) chiral Schiff base complexes. *J. Mol. Catal. A Chem.* 130: 41–50.

147. Kholdeeva, O. A. and Vanina, M. P. 2001. Comparative study of aerobic alkene epoxidations catalyzed by optically active manganese(III) and cobalt(II) salen complexes. *Kinet Catal. Lett.* 73: 83–89.

148. Egami, H., Oguma, T., and Katsuki, T. 2010. Oxidation catalysis of Nb(salan) complexes: asymmetric epoxidation of allylic alcohols using aqueous hydrogen peroxide as an oxidant. *J. Am. Chem. Soc.* 132: 5886–5895.

149. Chu, Y., Liu, X., Hu, X. et al. 2012. Asymmetric catalytic epoxidation of α,β-unsaturated carbonyl compounds with hydrogen peroxide: additive-free and wide substrate scope. *Chem. Sci.* 3: 1996–2000.

3 Transition Metal-Catalyzed Asymmetric Sulfoxidations

The sulfinyl group is one of the most efficient and versatile chiral controllers in C-C and C-X bond formations [1–6]. Chiral sulfoxides act as convenient auxiliaries in asymmetric synthesis. Many biologically active sulfoxides have been approved as drugs, for example, (S)-omeprazole (AstraZeneca's Nexium™ inhibitor of gastric acid secretion [7–10]) and (R)-lansoprazole (Takeda's Kapidex™ and Dexilant [11,12] anti-ulcer agents; (R)-modafinil [13,14], a stimulant-like drug; Sulindac [15], an anti-inflammatory; and others [16–21] (Figure 3.1).

Enantioselective synthesis of sulfoxides by transition metal-mediated asymmetric oxidation of sulfides was pioneered by the Kagan [22,23] and Modena [24] groups. They successfully adapted the Sharpless titanium tartrate–*tert*-butyl hydroperoxide asymmetric olefin epoxidation system [25] for enantioselective sulfoxidations. Major drawbacks of Kagan-Modena catalyst systems are their low productivity (sometimes nearly stoichiometric catalyst loads were required), complexity, the need for precise moisture control, and the use of alkyl hydroperoxides as oxidants.

Thus, they are not preferred oxidants from safety (flammability, corrosivity, hazardous effects) and atom economy (production of high molecular weight organic wastes) views. Nevertheless, most practical applications to date have exploited various versions of Kagan-Modena systems due to their broad substrate scope and ease of implementation.

Typically, older time-proven catalytic systems are preferred over challenging new developments by manufacturers that dominate the market. Hopefully, the continuing progress and environmental concerns of the community will reverse this situation in the future.

Since the milestone works of Kagan and of Modena, many metal-based catalyst systems for asymmetric sulfoxidations appeared [16–21]. Several successfully use hydrogen peroxide as a terminal oxidant. Only a few examples utilize molecular oxygen with [26,27] or without co-reductants [28]. An overview of asymmetric sulfoxidations with H_2O_2 on chiral metal complexes is given below.

VANADIUM SYSTEMS

To date, vanadium occupies an important place among enantioselective sulfoxidations with H_2O_2. The story began in 1995, when Bolm and Bienewald reported the catalytic properties of in situ–formed vanadium catalysts (from vanadyl(IV) acetyl acetonate and 1.5 equivalents of Schiff base ligand of type **1a**, **1b**, and **1c** (Figure 3.2) [29]. The catalyst system demonstrated high efficiency (so that 1.0 to 0.01 mol% vanadium load could be used) and oxidant economy: H_2O_2 was taken in only a

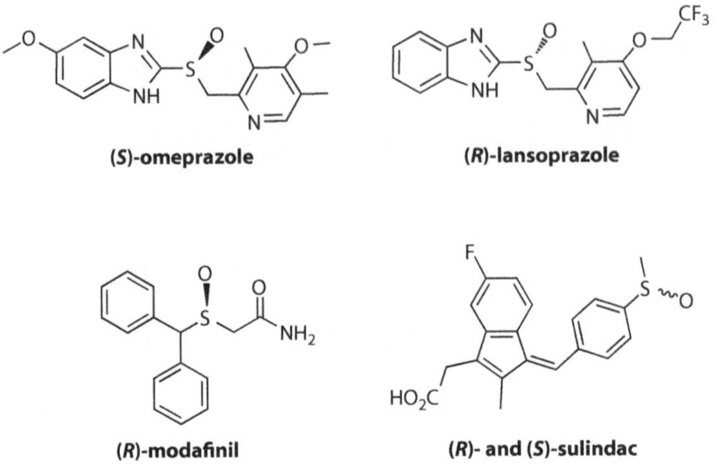

(S)-omeprazole

(R)-lansoprazole

(R)-modafinil

(R)- and (S)-sulindac

FIGURE 3.1   Chiral sulfoxides: biologically active compounds.

1.1-fold excess. Using 1 mol% of the catalyst, moderate to good enantioselectivities were reported (53 to 76% *ee;* Table 3.1) that approached 85% *ee* for the oxidation of 2-phenyl-1,3-dithiolane to the corresponding monosulfoxide [29,30].

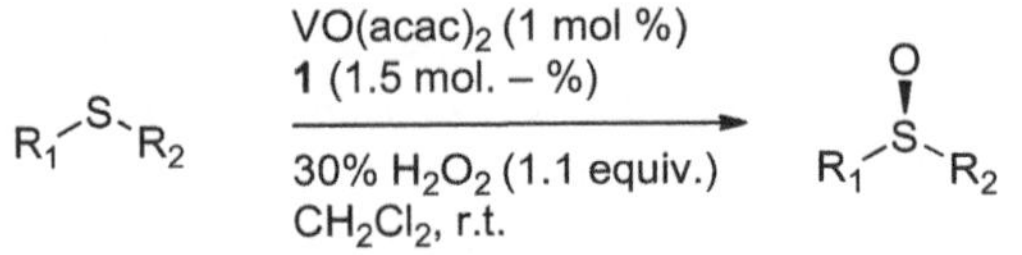

Later, the authors examined the scope of their catalyst system in the oxidation of a variety of dithioacetals and dithioketals [31]. On the basis of experimental observations such as ligand-accelerated reaction and proportional dependence of ligand optical purity and enantioselectivity, the authors predicted that the active sites should contain one chiral ligand per vanadium [29,30]. On the other hand, formation of several vanadium(V) species upon the addition of H_2O_2 was reported on the basis of ^{51}V NMR data [31]. Bryliakov and co-workers identified two major monoperoxovanadium(V) complexes (existing in equilibrium at about 1:1 ratio) by multinuclear NMR spectroscopy, and confirmed their reactivities toward sulfides [32,33]. A discussion of the structure of the catalyst was continued by Ellman [34], Maeda [35], and others [36–39].

The milestone work of Bolm and Bienewald inspired extensive research. Skarzewski and co-workers found that ligand **1d** was most effective for the asymmetric oxidation of various *bis*(arylthio)alkanes into the corresponding mono- and bis-sulfoxides [40,41]. Vetter and Berkessel synthesized a series of *tert*-leucinol-derived ligands **2-4** (Figure 3.2) bearing additional elements of chirality (central, planar, and axial) in the aldehyde moiety; ligand **2** showed the best enantioselectivity in the series in the oxidation of thioanisole: 78% *ee* at 0°C [42].

Katsuki et al. focused on the asymmetric oxidation of aryl methyl sulfides using a variety of similar axially chiral ligands. The highest enantioselectivity they reported was achieved with methyl 2-naphthyl sulfide (93% *ee* at 0°C) and ligand **5** [43].

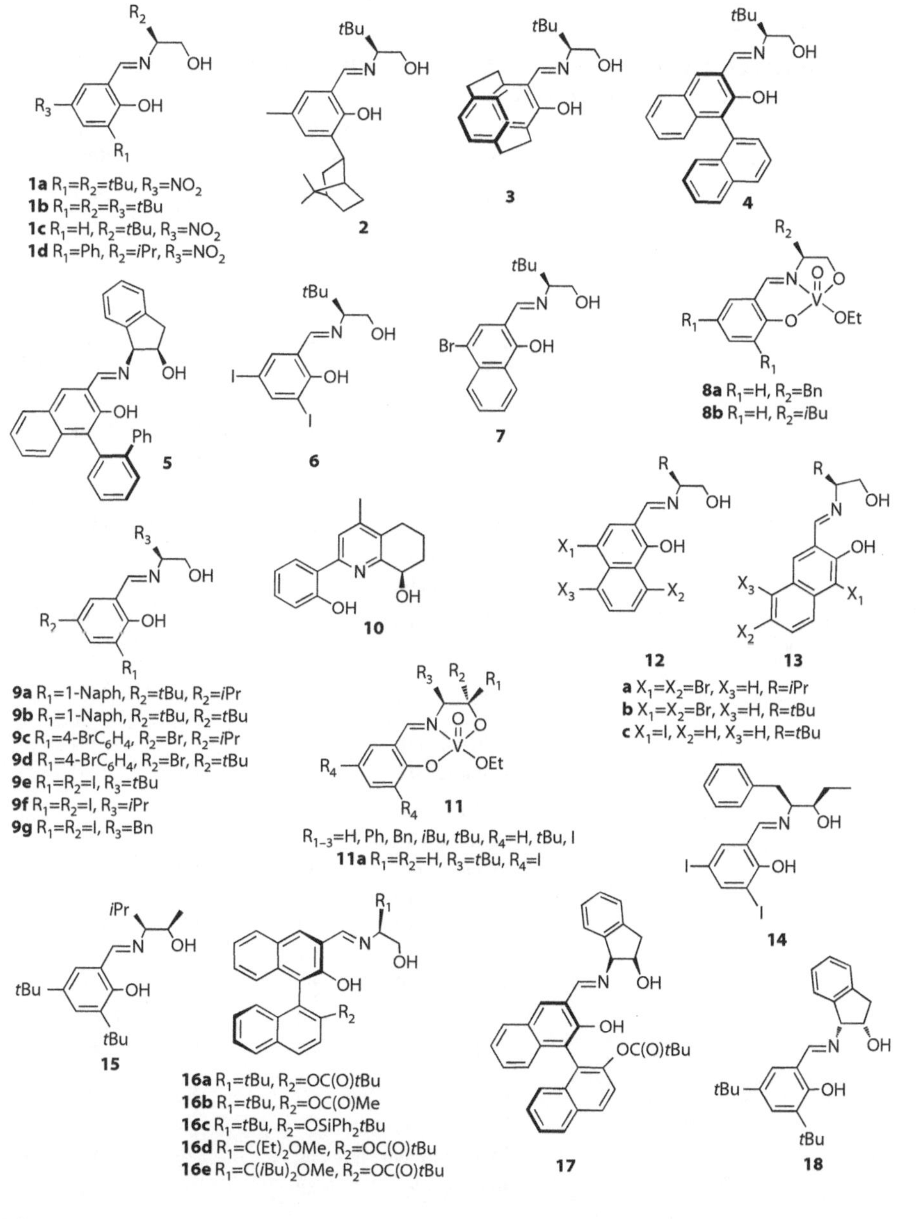

FIGURE 3.2 Schiff base ligands used as chiral auxiliaries in vanadium-catalyzed oxidations with H_2O_2.

Ellman's group reported the vanadium-catalyzed asymmetric mono-oxygenation of *tert*-butyl disulfide using ligand **1b** with up to 91% ee [44]. Somanathan and co-workers synthesized a library of Schiff base ligands from various salicylaldehydes and chiral β-amino alcohols, and reported moderate enantioselectivities (up to 65% ee) in vanadium-catalyzed oxidation of sulfides to sulfoxides [45,46].

TABLE 3.1
Enantioselective Oxidation of Sulfides with H_2O_2 in Presence of Vanadium–Schiff Base Catalysts

N	Substrate	Catalyst	Sulfoxide Yield (%)	ee (%)	Ref.
1		VO(acac)$_2$/**1a**	94	70	[29]
		VO(acac)$_2$/**1d**	90	75	[40]
		VO(acac)$_2$/**2**	97	78	[42]
		VO(acac)$_2$/**6**	81	90	[47]
		VO(acac)$_2$/**6**	70	96.7	[49]
		8a	65	86	[50]
		VO(acac)$_2$/**9d**	82	77	[54]
		11a	61	98	[57]
		VO(acac)$_2$/**12b**	57	94	[59]
		VO(acac)$_2$/**12c**	62	98	[59]
		VO(acac)$_2$/**14**	81	99	[60]
2		VO(acac)$_2$/**7**	74	90	[47]
		VO(acac)$_2$/**6**	76	97.4	[50]
		VO(acac)$_2$/**14**	84	88	[60]
3		VO(acac)$_2$/**1b**	64	62	[29]
4	Ph~S~~S~-Ph	VO(acac)$_2$/**1d**	60[a]	95	[40]
5	tBu~S~S~tBu	VO(acac)$_2$/**1b**	98[b]	91	[44]
		VO(acac)$_2$/**18**	90	85	[64]
6		VO(acac)$_2$/**6**	78	97	[47]
		VO(acac)$_2$/**6**	73	>99.5	[49]
		VO(acac)$_2$/**9d**	85	90	[54]
		VO(acac)$_2$/**10**	92	77	[56]
		VO(acac)$_2$/**12b**	67	98	[59]
		VO(acac)$_2$/**12c**	61	99	[59]
		VO(acac)$_2$/**14**	85	99	[60]
7		VO(acac)$_2$/**6**	70	>99.5	[49]
		VO(acac)$_2$/**14**	85	99	[60]
8		**8a**	59	86	[50]
		VO(acac)$_2$/**6**	51	91	[52]
		VO(acac)$_2$/**16d**	80	96	[62]
9		VO(acac)$_2$/**6**	48	94	[52]
10		VO(acac)$_2$/**1b**	84[c]	85	[29]
		VO(acac)$_2$/**1d**	80	88	[40]
		VO(acac)$_2$/**15**	88	>99.9	[61]

[a] (R,R)-*bis*-sulfoxide yield.
[b] Sulfide conversion reported.
[c] Monooxide yield.

Anson and Jackson with co-workers screened a library of Schiff bases derived from natural and synthetic β-amino alcohols and aromatic o-hydroxy aldehydes, and chose ligands **6** and **7** with halogen substituents in aromatic rings. They demonstrated the highest enantioselectivities in the vanadium-catalyzed enantioselective oxidation of sulfides with H_2O_2 in dichloromethane, yielding oxides of alkyl aryl sulfides at 74 to 86% yield and 89 to 97% ee [47].

In a subsequent work, vanadium Schiff base systems (with ligand **6** and its enantiomer) catalyzed the oxidative kinetic resolution of racemic sulfoxides. Higher stereoselectivity factor S (ratio of rate constant of fast-reacting and slow-reacting enantiomers) values were found in $CHCl_3$ (10.5 to >30 at 0°C), being generally lower in toluene [48]. Those findings opened a door for designing a combined stepwise vanadium-catalyzed asymmetric oxidation and kinetic resolution procedure that was exploited to obtain alkyl aryl sulfoxides in over 70% yield and up to 99.5% ee (using ligand **6** or its enantiomer) [49].

In an independent work, Zeng and co-workers prepared vanadium complexes of type **8** and studied the oxidation of aryl methyl sulfides and benzyl phenyl sulfide in CH_2Cl_2 [50]. They reported good enantioselectivities (76 to 99% ee) and generally moderate yields (31 to 78%) and the crucial role of kinetic resolution for the formation of sulfoxides with high-optical purity. The phenomenon of kinetic resolution of sulfoxides was also studied and exploited in the works of Maguire's group who focused on the synthesis of aryl benzyl sulfoxides from the corresponding sulfides [51,52]. They screened a series of Schiff base ligands and identified **6** as the most efficient chiral inducer. In the combined asymmetric oxidation and kinetic resolution procedure, a series of bulky sulfoxides were synthesized to 56 to 99% ee, with yields not exceeding 54% [51,52]. The efficiency of the kinetic resolution was substrate- and solvent-dependent; the highest ees were achieved for substituted benzyl aryl sulfoxides in CH_2Cl_2 [52].

Zeng with co-workers prepared a set of vanadium-Schiff base complexes and tested them as catalysts in the oxidation of allyl and cinnamyl aryl sulfides with H_2O_2. Using the catalyst derived from ligand **6**, various substrates were oxidized to the corresponding sulfoxides with moderate yields (43 to 78%) and 33.5 to 97.3% ee [53].

Sun and co-workers tested a series of 3-aryl-substituted Schiff bases of the general formula **9**; some appeared rather effective chiral auxiliaries in vanadium-catalyzed oxidations of aryl methyl sulfides (53 to 92% ee) [54,55]. Structures featuring tert-butyl substituent at the asymmetric carbon in all cases demonstrated higher asymmetric induction compared to those with iso-propyl and benzyl. Subsequently, they reported a tridentate ligand **10** that showed enantioselectivities ranging from 8 to 77% ee in asymmetric oxidations of several aryl methyl sulfides in acetone, with good to high sulfoxide yields (80 to 95%) [56].

Gau et al. isolated and characterized via x-rays a series of vanadium(V) complexes of type **11** and used them as pre-catalysts for asymmetric oxidation of methyl phenyl sulfide with H_2O_2 [57]. Moderate to high yields (61 to 80%) and ees (26 to 98%) were reported. The enantioselectivity increased at low temperatures, and the use of mixed CH_2Cl_2/toluene solvent instead of CH_2Cl_2 afforded methyl phenyl sulfoxide up to 98% ee (with catalyst **11a**) at the expense of lowered yield (61%).

Apparently, the effect of kinetic resolution of sulfoxides is more pronounced upon the addition of toluene.

Further progress with vanadium Schiff catalyzed sulfoxidations was achieved mostly via ligand-oriented design of novel catalysts. Pati and co-workers developed a series of trimeric Schiff bases with various amino alcohol moieties and a central linker. Using 0.6 mol% of their ligands and 1.0 mol% of VO(acac)$_2$ as the catalyst, they reported the synthesis of benzyl phenyl sulfoxide at 92 to 98% yield and up to 89% *ee* [58].

Wang and Sun with co-workers presented a new family of tridentate ligands of types **12** and **13** derived from β-amino alcohols and bromo- and iodo-functionalized hydroxynaphthaldehydes [59]. A series of aryl methyl sulfides were oxidized with up to 99% *ee* and moderate yields (50 to 60%) in toluene. In CH$_2$Cl$_2$, the sulfoxide yields were higher (80 to 93%), albeit with lower enantioselectivities. The authors claimed the results to have industrial potential; however, the observed yields were rather low (51 to 67%), apparently due to an accompanying kinetic resolution of the resulting sulfoxides that was more pronounced in toluene [59].

Li and co-workers focused their research on tridentate Schiff bases bearing two stereogenic centers in the amino alcohol moieties [60]. One of the resulting ligands (**14**) appeared to be a very efficient chiral inducer for the vanadium-catalyzed sulfoxidations with H$_2$O$_2$. In chloroform, a series of aryl methyl sulfides were oxidized, producing 80 to 85% yield and 98 to 99% *ee* at 0°C [60]. High *ee* values were achieved in a tandem enantioselective oxidation and kinetic resolution process; this required 1.35 equivalents of H$_2$O$_2$ with respect to the substrate. The enantioselectivities reported by the authors are among the best for the asymmetric oxidation of alkyl aryl sulfides with H$_2$O$_2$.

Whether the substrate scope of this catalyst system is broader than simple alkyl aryl sulfides and whether the process may be extended to practical syntheses of biologically active chiral sulfoxides or their precursors remain challenges. The major shortcoming of the proposed reaction conditions is the low reactivity that requires 24- to 48-hr reaction time). Another ligand with two stereogenic centers (**15**) was identified as most efficient in the oxidation of 1,3-dithianes with H$_2$O$_2$ [61].

Ahn and co-workers concentrated on the synthesis of sterically hindered BINOL-derived Schiff base ligands of types **16** and **17** possessing additional elements of axial chirality [62,63]. Ligand **16a** showed high enantioselectivities in vanadium-catalyzed oxidation of thioanisole (90% yield, 86% *ee*) and up to 98 to 99% *ee* in the oxidation of benzyl phenyl sulfide and benzyl *p*-bromophenyl sulfide [62]. Ligand **17** led to a similar catalytic performance [63]. The authors noted that the absolute configuration of the sulfoxide was dictated by the chirality of the imine moiety.

Khiar and Fernandez with co-workers studied vanadium-catalyzed monooxidation of disulfides to enantiomerically pure thiosulfinates [64]. The best vanadium catalyst (VO(acac)$_2$; **18**) promoted the oxidation of *t*-butyl disulfide with 85% *ee*; the result is inferior to those of the original Ellman method [44]. A number of Schiff base ligands featuring carbohydrate-derived chiral diamine moieties were tested in asymmetric sulfoxidations and exhibited generally low *ee*s (0 to 45% and 60% in one case) [65,66]. Several tridentate Schiff base ligands were synthesized by condensation of various salicylaldehydes with chiral amino alcohols derived from α-pinenes

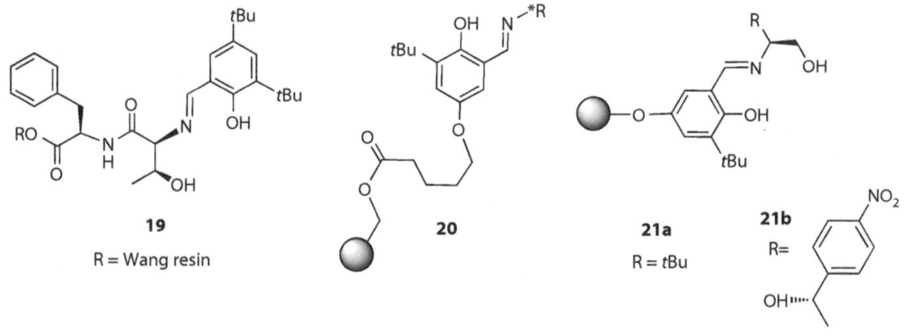

FIGURE 3.3 Immobilized Schiff base ligands.

and 3-carene; the enantioselectivities of vanadium-catalyzed sulfoxidations were low (1 to 32% *ee*) [67,68].

Several attempts to design immobilized vanadium Schiff base catalysts produced limited success. Jackson's group screened 72 chiral ligands fixed on a Wang resin; even the best of their supported ligands (**19**; Figure 3.3) oxidized methyl phenyl sulfide with only 23% *ee* [69]. Anson and Jackson screened a library of 29 Schiff bases of type **20** supported on Merrifield resin; the enantioselectivities in the oxidation of methyl phenyl sulfide with H_2O_2 were low (4 to 21% *ee* with the best chiral ligand featuring *tert*-leucinol moiety R*) [47]. Barbarini and co-workers attached a series of Schiff base ligands of type **21** to polystyrene or polyacrylate support; the oxidation of alkyl aryl sulfides was less stereoselective results from their homogeneous counterparts (61% *ee* for methyl phenyl sulfide) [70].

Reports of vanadium catalyst systems using other chiral ligands are rather scanty. Bryliakov and Talsi tested a series of salan ligands in vanadium-catalyzed oxidation of sulfides with H_2O_2 with low enantioselectivities (not exceeding 37.5% *ee*) [71]. Correia and Pessoa with co-workers isolated and characterized several $V^{IV}O$(salan) and $V^{IV}O$(salen) complexes that catalyzed the enantioselective oxidation of methyl phenyl sulfide with H_2O_2 with 3 to 47% *ee* [72].

Overall, vanadium–tridentate Schiff base–H_2O_2 catalyst systems displayed promising results in enantioselective oxidations of simple alkyl aryl sulfides and disulfides with various structures, and seem to have high potential. There have been attempts to use them for preparative enantioselective syntheses of biologically active sulfoxides [18,73,74]. In some cases, they demonstrated better chemo- and enantioselectivities as compared to the Kagan-Modena systems (e.g., for 2-aryl-1,3-dithiane and disulfide oxidations). Catalyst systems of this type are active and efficient and require low catalyst loads (typically 1 mol%). Other advantages of such systems are ease of handling (open vessels, often room temperature, no moisture control required), and the use of low cost and safe oxidants. A disadvantage is the limited substrate scope. For example, the oxidation of bulky sulfides (like aryl benzyl sulfides) with good yields and enantioselectivities at the same time remains a technical problem and tolerance to various functional groups remains poorly explored.

TITANIUM SYSTEMS

Historically, titanium occupies the most important position in asymmetric sulfoxidations, since most practical applications are based on Kagan-Modena systems, using alkyl hydroperoxides as oxidants. At the same time, titanium-catalyzed asymmetric oxidations with H_2O_2 were relatively rare until the 2000s. To date, several titanium–H_2O_2 catalyst systems have been reported; all of them use chiral Schiff bases or salan type chiral ligands.

The earliest example of this type was the titanium(IV)-Schiff base complex **22** (Figure 3.4) reported by Pasini et al. in 1986 [75]. Catalyst **22** demonstrated excellent efficiencies (1000 to 1500 TON) in the oxidation of methyl phenyl sulfide by H_2O_2 in aqueous methanol or dichloromethane, but the optical purity of the resulting sulfoxide was rather low (<20% ee). Katsuki and Saito later tested the mononuclear "second generation" titanium salen complex **23**, bearing additional elements of axial chirality, as a catalyst for the oxidation of methyl phenyl sulfide with aqueous hydrogen peroxide and reported poor enantioselectivity [76].

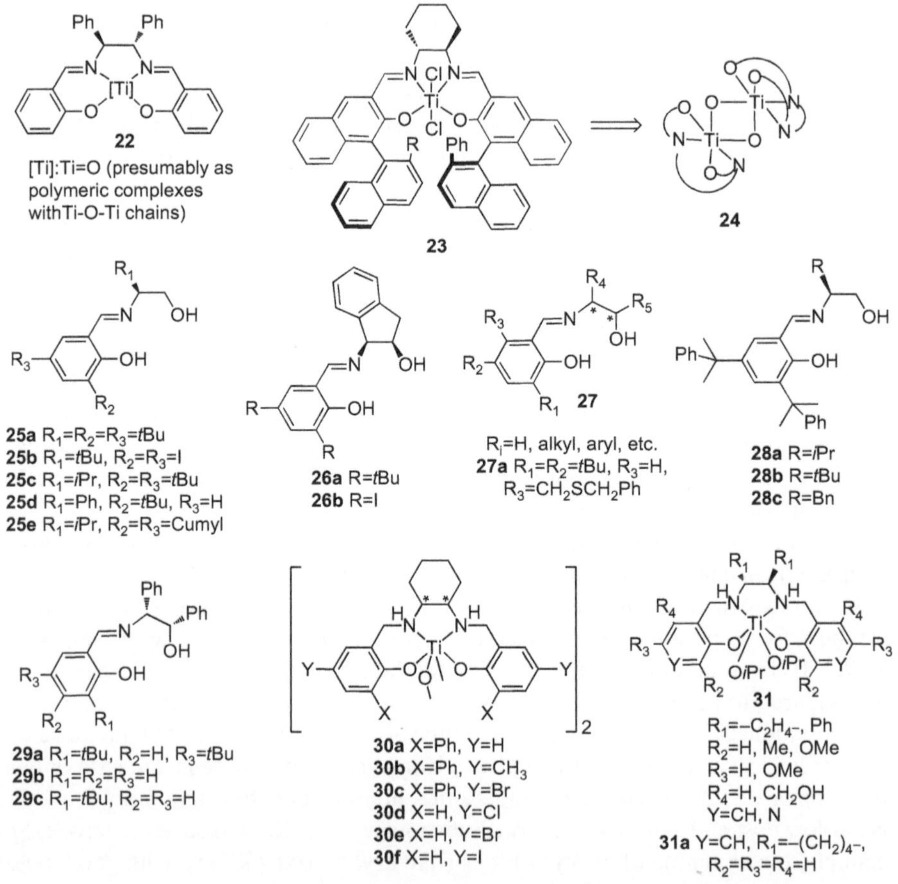

FIGURE 3.4   Titanium catalysts and ligands for asymmetric sulfoxidations with H_2O_2.

Surprisingly, when **23** was converted to the corresponding bis-μ-oxo binuclear titanium complex **24**, the latter (taken in 2 mol%) demonstrated high activity and selectivity in the oxidation of small alkyl aryl sulfides with H_2O_2 or with urea hydroperoxide (UHP) in methanolic solution (Table 3.2). The authors discussed possible reaction mechanisms and proposed that the most likely oxidizing agent could be a mononuclear titanium peroxo complex [77]. Subsequently, it was reported that the system **24**–UHP could be remarkably selective for the asymmetric monooxidation of cyclic dithioacetals (with *ees* of the resulting mono-sulfoxides ranging from 39 to 99%) [78].

$$Ar^{\diagdown S \diagdown} Me \quad \xrightarrow[\text{MeOH, 0 °C, 24h}]{\textbf{24} \text{ (2 mol \%), UHP(1 eq.)}} \quad Ar^{\diagdown \overset{\overset{O}{\|}}{S} \diagdown} Me$$

Another approach was chosen by Jackson and co-workers who tested an immobilized (Wang resin-supported) Schiff base ligand **19** for titanium-catalyzed sulfoxidations [69]. Upon the use of 1.5 mol% of the solid catalyst CH_2Cl_2 and 1.1 equivalents of H_2O_2, the authors reported moderate enantioselectivities (45 to 72% *ee* in CH_2Cl_2). Bryliakov and Talsi synthesized a new family of titanium-based catalysts with *N*-salicylidene-*L*-amino alcohols of types **25** and **26**, capable of asymmetric oxidation of prochiral sulfides with H_2O_2 [79,80]. The reactions proceeded with high selectivity and conversion levels, but with moderate enantioselectivities, not exceeding 65% *ee* for benzyl phenyl sulfide [80].

$$R_1^{\diagdown S \diagdown} R_2 \quad \xrightarrow[\substack{H_2O_2 \text{ (30\%)} \\ CH_2Cl_2}]{\textbf{19}/\text{Ti (O}i\text{Pr)}_4} \quad R_1^{\diagdown \overset{\overset{O}{\|}}{S} \diagdown} R_2$$

$$R_1 = \text{Aryl, } R_2 = \text{Me, Et, Pr}$$

The effect of kinetic resolution was negligible [79]. Titanium–tridentate Schiff base systems appeared to be more active than systems based on analogous vanadium (see above) and iron (see below). Oxidations were performed successfully within hours at catalyst and substrate concentrations one order of magnitude lower than in Fe and V systems. As distinct from vanadium(V) systems, the replacement of the *t*Bu substituent in the amine moiety with *i*Pr led to higher *ees*.

An NMR study provided evidence for the existence of titanium complexes as mononuclear species bearing one chiral ligand per titanium center. Somanathan with co-workers prepared a set of chiral β-amino alcohol-derived Schiff bases of type **27**, bearing two asymmetric centers at the amino alcohol moiety [46]. Titanium-catalyzed asymmetric oxidation of methyl phenyl sulfide with H_2O_2 in the presence of **27** demonstrated slightly higher enantioselectivities (up to 64% *ee*) compared to results from vanadium counterparts under the same conditions.

Wang and Sun et al. prepared a series of dicymenyl-substituted amino alcohol-derived Schiff bases **28** with various substituents at the stereogenic carbon and tested them in titanium-catalyzed oxidations of aryl methyl sulfides [81]. From a series of

TABLE 3.2

Enantioselective Oxidation of Sulfides in Presence of Titanium Catalysts

N	Substrate	Catalyst	Oxidant	Sulfoxide Yield (%)	ee (%)	Ref.
1		19/Ti(O*i*Pr)$_4$	H$_2$O$_2$	100[a]	64	[69]
		27a/Ti(O*i*Pr)$_4$	H$_2$O$_2$	78	64	[46]
		28b/Ti(O*i*Pr)$_4$	H$_2$O$_2$	89	73	[81]
		29a/Ti(O*i*Pr)$_4$	H$_2$O$_2$	85	84	[82]
		30f	H$_2$O$_2$	78	77.5	[85]
2		24	UHP	91	93	[76]
		19/Ti(O*i*Pr)$_4$	H$_2$O$_2$	96[a]	64	[69]
3		30f	H$_2$O$_2$	72	79.5	[85]
4		30f	H$_2$O$_2$	80.5	67.5	[85]
5		24	UHP	93	96	[76]
		28b/Ti(O*i*Pr)$_4$	H$_2$O$_2$	79	59	[81]
		29a/Ti(O*i*Pr)$_4$	H$_2$O$_2$	78	82	[82]
6		29a/Ti(O*i*Pr)$_4$	H$_2$O$_2$	49	98	[82]
7		30e	H$_2$O$_2$	75	82	[85]
8		19/Ti(O*i*Pr)$_4$	H$_2$O$_2$	87[a]	72	[69]
9		25c/Ti(O*i*Pr)$_4$	H$_2$O$_2$	83	32	[79]
		25c/Ti(O*i*Pr)$_4$	H$_2$O$_2$	86	60	[79]
		25e/Ti(O*i*Pr)$_4$	H$_2$O$_2$	85	65	[80]
		29a/Ti(O*i*Pr)$_4$	H$_2$O$_2$	78	65	[82]
		30a	H$_2$O$_2$	73	98.5	[84]
		30f	H$_2$O$_2$	81	92.5	[85]
10		30a	H$_2$O$_2$	75	96.5	[84]
11		24	UHP	91	99	[78]

[a] Sulfide conversion reported.

catalytic experiments, ligand **28b** emerged as the most efficient chiral inducer, yielding 89% methyl phenyl sulfoxide and 73% *ee*. Ligands **28a** and **28c** (with *i*Pr and Bn) substituents at the stereogenic carbon led to lower *ee*s. The effect of reaction media was extensively studied; CH_2Cl_2 was found the best solvent for achieving good enantioselectivity and conversion at the same time.

Abdi with co-workers prepared a series of ligands of type **29** with two stereogenic centers at the amino alcohol moiety [82]. Chiral ligand **29a** demonstrated the highest asymmetric induction in the series; in one case, enantioselectivity up to 98% *ee* was reported. The ligand structures reported appear to be the most productive for the oxidation of small aryl methyl sulfides; more sterically demanding substrates (PhSEt, $PhSCH_2Ph$) were oxidized with poorer enantioselectivities. The authors reported apparent first orders in the catalyst and substrate concentrations and, unexpectedly, zero order in the oxidant, which is likely to be an artifact [83]. Indeed, saturation behavior was previously documented for the dependence of apparent sulfoxidation rate constant at H_2O_2 concentrations of 0.3 to 0.4 M [79]. At lower H_2O_2 concentrations, the observed rate constant increased with raising concentration of added oxidant.

Possible reasons for the saturation behavior could be (1) conversion of all titanium into the peroxotitanium intermediate at high H_2O_2 concentrations, (2) saturation of the (dichloromethane) reaction mixture in H_2O_2, or both. Summarizing these observations, the most likely proposed reaction scheme may explain the observed dependences of the sulfoxidation rate.

Another promising class of asymmetric sulfoxidation catalysts is based on titanium–salan complexes, also used by Katsuki et al. for asymmetric epoxidations with H_2O_2. Bryliakov and Talsi examined a series of dinuclear titanium-salan complexes of type **30** as catalysts for the enantioselective oxidation of prochiral sulfides with H_2O_2 [71]. The titanium complexes could perform at least 500 turnovers with no loss of enantioselectivity. The kinetic resolution of the resulting sulfoxides substantially improved the optical purity of the sulfoxide products at high conversions, to achieve up to 98.5% *ee* [71,84].

Later, the oxidation of bulky sulfides (basically substituted benzyl aryl sulfides) was studied in more detail [84]. The general structure of catalysts **30a–30c** was most efficient for the oxidation of bulky (aryl benzyl) sulfides. The introduction of electron-donating or electron-withdrawing substituents at the fifth positions (**30b**, **30c**) deteriorated the outcome. At the same time, the oxidation of smaller sulfides with high enantioselectivities remained a problem with catalysts **30a–30c**. However, this problem could be resolved by the use of catalysts **30d–30f** [85]. Moreover, the latter are capable of catalyzing the oxidation of bulky sulfides with enantioselectivities comparable to **30a**. A proper choice of experimental conditions allowed reasonably high yields of sulfoxides (77 to 87%) and good to high *ees* at the same time [84,85].

Pessoa and Correia with co-workers prepared a series of chiral titanium–salan complexes of type **31** [86] and reported low to moderate enantioselectivities (up to 51% *ee* with **31a**) for the oxidation of thioanisole with H_2O_2, with 16 to 71% sulfoxide yields. They tested their catalysts in various reaction media (acetone, ethyl acetate, ionic liquids). The highest enantioselectivities were documented in dichloroethane. Barman et al. synthesized two helical titanium–salan catalysts that showed low enantioselectivities in thioanisole oxidation (up to 43% *ee*) [87].

To date only indirect data describe the nature of active species responsible for enantioselective oxygenations in these catalyst systems. In particular, the groups of Katsuki and of Berkessel concluded that the elusive active epoxidizing species in titanium salan- and salalen-based catalyst systems were most likely mononuclear species (see Chapter 2).

Talsi and Bryliakov examined the kinetics of parallel oxidations of various sulfides on titanium–salan catalysts and found that the first (rate-determining) step is H_2O_2 interaction with the Ti catalyst to form the active oxidant [85]; the latter further rapidly reacts with the sulfide. The authors carried out several competitive oxidation experiments (with various *p*-substituted thioanisoles) that revealed the moderately electrophilic nature of the active species responsible for the enantioselective oxygen transfer, with close Hammett slope ρ values for four different catalysts, falling in the range of −1.35 to −1.40. These data are consistent with a concerted mechanism in which an electrophilic oxygen-transferring species is directly attacked by a nucleophilic sulfide, while an alternative electron transfer mechanism (with formation of a transient sulfenium radical cation intermediate) can be ruled out [85].

The following sequence can be proposed. At the first (rate-determining) step, an adduct is formed between the titanium catalyst and H_2O_2. Coordination of H_2O_2 to the titanium center in the chiral ligand framework is believed to increase the peroxide electrophilicity and be a precondition to accomplish the oxygen transfer in a stereoselective fashion. A similar peroxotitanium reactive intermediate was predicted

active species

by DFT calculations and its formation was corroborated by atmospheric pressure chemical ionization mass spectrometry studies [88]. The presence of the N-H moiety is essential for efficient catalysis (see above and Refs. [71] and [87]). Apparently, hydrogen bonding between the N-H group and peroxo oxygen assists in activating the peroxo group for oxygen transfer.

In comparison to the classical Kagan-Modena Ti-diethyltartrate–alkyl hydroperoxide sulfoxidations, modern titanium-based catalyst systems are more efficient (performing 50 to 500 turnovers), environmentally safe, and atom-economic. Catalyst systems based on chiral β-amino alcohol-derived Schiff bases developed to date demonstrate good turnover (up to 100), good sulfoxide selectivities (>90%), and moderate to good enantioselectivities (generally 50 to 70% and up to 98% *ee* in one case). However, their substrate scope is mostly limited to small aryl methyl sulfides, while bulky sulfides (potential models of precursors of real biologically active sulfoxides) are oxidized with only moderate stereoselectivities—insufficient for practical applications. Titanium–salan catalysts are more efficient (sustaining 500 or more turnovers) and allow the oxidation of bulky sulfides with reasonably good yields and good to excellent enantioselectivities. Hopefully, catalyst systems of such types can find applications in preparative asymmetric syntheses of biologically active sulfoxides or their precursors.

IRON SYSTEMS

Apparently, the first iron-based catalyst system capable of catalyzing enantioselective sulfoxidations with H_2O_2 was reported in 1997 by Fontecave and co-workers who prepared a model binuclear non-heme iron(III) complex **32** (Figure 3.5) [89]. The latter afforded *p*-bromophenyl methyl sulfoxide in 40% *ee* and 90% yield (based on the oxidant; the system was tested at the iron:substrate:oxidant ratio of 1:600:10). The authors identified the active species as a peroxo adduct of **32** and concluded that the reaction proceeded through a nucleophilic attack of the sulfide to the peroxoiron intermediate [90]. A possible explanation of the synergistic effect of two iron sites was peroxide coordination to one iron site and sulfide coordinated to the other [91].

In 2003, Bolm and Legros attempted to apply the β-amino alcohol-derived tridentate chiral Schiff bases **33** (earlier introduced in vanadium-catalyzed asymmetric sulfoxidations (see above)) for iron-catalyzed sulfoxidations. Using 4 mol% of the chiral ligand and Fe(acac)₃, and 1.2 equivalents of H_2O_2, a series of aryl methyl sulfides were oxidized to sulfoxides with 13 to 90% *ee* in generally low yields (15 to 44%) (Table 3.3); ligand **33e** displayed the highest enantioselectivity [92].

Bulkier sulfides (PhSEt, PhSBn) were oxidized with lower enantioselectivities. Later, the catalytic performance was improved by the introduction of appropriate catalytic additives (substituted benzoic acids or their lithium salts); compounds of type **34** (0.5 equivalent with respect to iron) was the best additive [93]. In effect, both the yields of sulfoxides and their optical purities improved significantly (up to 95 or 96% *ee* in some cases). The mechanistic rationale for the use of 0.5 equivalent of additive is not entirely clear; the authors supposed that monocarboxylate-bridged di-iron(III) complexes could be involved [93].

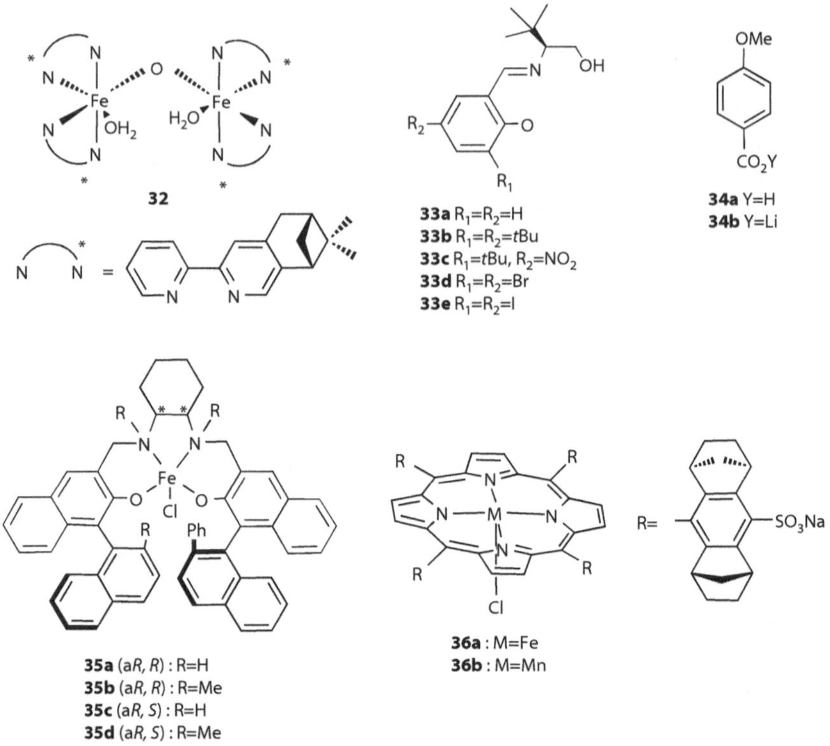

FIGURE 3.5 Iron-based catalysts and ligands for asymmetric sulfoxidations with H_2O_2.

In a subsequent work, a broader range of catalytic additives was considered [94]. The authors did not observe any effect of kinetic resolution on the sulfoxidation enantioselectivity; possible participation of more than one chiral ligand at the stereochemistry-determining step on the basis of non-linear dependence of the sulfoxide *ee* on the ligand *ee* was proposed [94]. Both enantiomers of the Sulindac non-steroidal anti-inflammatory drug were prepared using this catalyst system with up to 92% *ee* and up to 71% yield [95].

Katsuki and Egami reported that chiral iron(III)–salan complexes of type **35** (in 1 mol%) efficiently catalyzed the oxidation of several aryl methyl sulfides to sulfoxides with H_2O_2, in generally high yields (76 to 99%) and enantioselectivities (81 to 96% *ee* with catalyst **35d**) [96]. The catalyst system required no additives or organic solvents; the reactions successfully proceeded in water at room temperature. In most cases, significant amounts of sulfone (7 to 24%) formed; the presence of *o*-substituent in the sulfide aryl ring effectively suppressed overoxidation.

In a subsequent publication, the authors reported a broadened substrate scope of their catalysts and noted that the reaction conditions could be optimized by lowering the temperature and reducing the catalyst load to 0.2 mol% [97]. Yang and co-workers encapsulated a chiral Fe(salan) complex **35c** in nanocages of modified mesoporous silicas. Several *p*-substituted aryl methyl sulfides were oxidized to sulfoxides with high conversions (85 to 98%) and selectivities (85 to 95%), but the

TABLE 3.3
Enantioselective Oxidation of Sulfides with H_2O_2 in Presence of Iron Catalysts

N	Substrate	Catalyst	Additive	Sulfoxide Yield (%)	ee (%)	Ref.
1		33e	—	36	59	[92]
		33e	71b	63	90	[93]
		35d	—	92	96	[96]
		35c/SiO$_2$[a]	—	95	80	[98]
		35c	—	89	87	[98]
		36a	N-Me-Imd	56	90	[99]
2		33e	—	30	44	[92]
		33e	71b	56	82	[71]
		35d	—	78	81	[86]
		35c/SiO$_2$[a]	—	81	73	
3		35d	—	91	88	[86]
4		33e	—	44	70	[92]
		33e	71a	67	95	[93]
5		36a	—	83	79	[99]
6		33e	—	40	27	[92]
		33e	71a	73	79	[93]
7		35d	—	97	96	[96]

[a] Fe catalyst was encapsulated in nanopores of modified silica.

enantioselectivities were 1 to 16% *ee* lower than those demonstrated by the homogeneous system under the same conditions [98].

In 2011, the first example of iron–porphyrin-catalyzed sulfoxidation with hydrogen peroxide was reported [99]. The authors used a chiral water-soluble porphyrin complex **36a** that, at 1 mol% load, catalyzed the oxidation of substituted thioanisoles with aqueous H_2O_2 in methanol (Table 3.3), demonstrating good yields (91 to 98%) and enantioselectivities (71 to 90% *ee*). The authors noted that the addition of *N*-methylimidazole resulted in an increase of reaction time and enantioselectivity [99].

Very few iron-based catalyst systems for asymmetric sulfoxidations with H_2O_2 are known to date; most demonstrate good efficiencies (with 1 to 4 mol% catalyst loads) and good to excellent enantioselectivities. However, their substrate scope is limited to aryl methyl sulfides; an exception is the system of Bolm and Legros [93]. Only the latter iron-based system has been applied successfully to synthesize the optically pure biologically active Sulindac sulfoxide [95].

SYSTEMS BASED ON OTHER METALS

Metals other than V, Ti, and Fe have been exploited rarely in asymmetric sulfoxidations with hydrogen peroxide. One of the most interesting catalyst systems is based on the aluminum–salalen complexes of type **37** proposed by Katsuki and co-workers (Figure 3.6) [100]. Catalyst **37d** (the most stereoselective of the series) was reported to catalyze the oxidation of substituted thioanisoles at 81 to 91% yield and 97 to

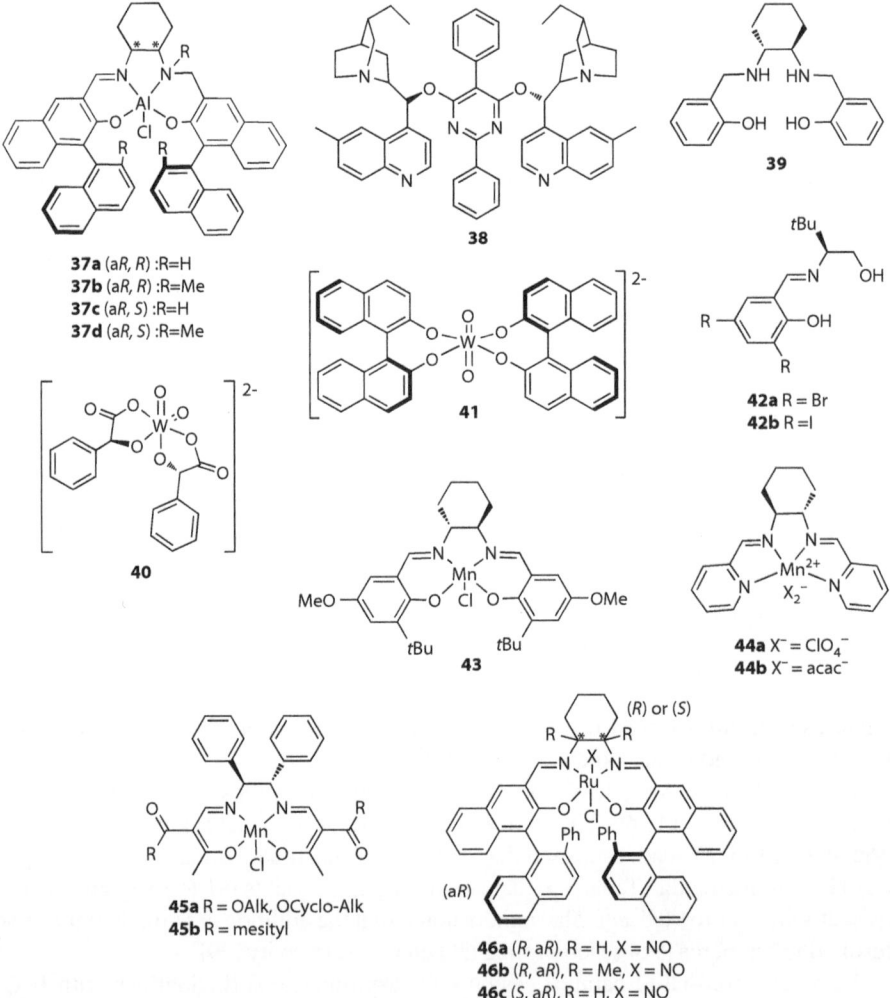

FIGURE 3.6 Catalysts based on other metals.

99% *ee* (Table 3.4). The reactions proceeded for 24 hr at room temperature and required 2 mol% catalyst loads.

Although the high enantioselectivities were due to a tandem asymmetric oxidation and kinetic resolution process, the reported yields of sulfone in the solvent of choice (methanol) did not exceed 10% [100]. Subsequently, the authors found that the same system could effectively operate under solvent-free conditions and at lowered catalyst loads (0.01 to 0.002 mol%), yielding sulfoxides with similarly high optical purities. The chemical yields were lower in some cases: 31 to 92%). The authors hypothesized that an aluminum η^2-hydroperoxo species could be the active

TABLE 3.4
Enantioselective Oxidation of Sulfides with H_2O_2 by Catalyst Systems Based on Other Metals

N	Substrate	Catalyst	Additive	Sulfoxide Yield (%)	ee (%)	Ref.
1		37d	—	86	98	[100]
		WO$_3$/(DHQD)$_2$–PYR[a]	—	88	59	[104]
		43	—	90	47	[113]
		46b	H$_2$O	74	94	[119]
2		37d	—	80	91	[100]
		WO$_3$/(DHQD)$_2$–PYR[a]	—	44	82	[104]
3	Ph–S–S–Ph	37d	—	31	91	[101]
4		45b	—	57	51	[117]
5		37d	—	82	98	[100]
		WO$_3$/39	—	86	58	[105]
		43	—	95	42	[113]
		45b	—	58	44	[117]
		46b	H$_2$O	61	91	[119]
6		WO$_3$/(DHQD)$_2$–PYR[a]	—	88	61	[104]
		Cu(acac)$_2$/40a	Ionic liquid	21	71	[107]
		Cu(acac)$_2$/40a		90	79	[108]
7		37d	—	92	98[b]	[102]
		46b	H$_2$O	57	98	[119]
8		45b	—	72	72	[117]
		46b	H$_2$O	99	94	[119]

[a] Cinchona alkaloid.
[b] Monosulfoxide yield.

oxidizing factor [101]. Catalysts of this type also proved capable of highly enantio- and diastereoselective asymmetric monooxidation of cyclic dithioacetals [102,103].

In 2003, Sudalai and Thakur reported a heterogeneous catalyst based on WO_3 and cinchona alkaloids that catalyzed the oxidation of several alkyl aryl sulfides with H_2O_2 to 90% *ee* [104]. The best results were obtained by using WO_3 (5 mol%) and $(DHQD)_2$-PYR **38** (10 mol%). The involvement of kinetic resolution of sulfoxides (resulting in higher *ee*s and lower yields) was established.

Nevertheless, the antiulcer sulfoxide drug (*R*)-lansoprazole was prepared in good yield (84%) and *ee* (88%) [104]. Zhang and Zhu with co-workers reported on a heterogeneous catalyst system based on WO_3–salan ligands. In combination with the best salan ligand **39** (5 mol% of tungsten and 5 mol% of ligand), they catalyzed the oxidation of aryl methyl sulfides with H_2O_2 at 27 to 67% *ee* [105]. Among other tungsten-based catalyst systems, chiral ionic liquids comprising chiral anions **40** and **41** can be mentioned. Thioanisole was oxidized to the corresponding sulfoxide with up to 95% *ee* with H_2O_2 and UHP, the latter demonstrating lower sulfoxide yields and higher enantioselectivities [106].

Maguire and co-workers reported on copper complexes with tridentate β-amino alcohol-derived Schiff base ligands that catalyzed the oxidation of aryl benzyl sulfides with H_2O_2 [107]. The catalysts were prepared in situ from $Cu(acac)_2$ (2 mol%) and a *t*-leucinol derived ligand (4 mol%) of type **42**. Chiral aryl benzyl sulfoxides were formed in CCl_4 with 14 to 49% isolated yield and 13 to 81% *ee* [107]. The use of catalytic additives, such as NMO, DMSO, or ionic liquid improved the enantioselectivity, and no sulfone formation was reported.

In a subsequent study, the authors formulated optimized reaction conditions (i.e., no additives and hexane–alcohol mixed reaction solvents), so that the reactions proceeded with much higher chemical yields (up to 91%) with no loss of chemoselectivity and equal or higher enantioselectivities [108].

Two modified β-cyclodextrin derivatives with catechol-type ligands were prepared and tested as ligands in Mo- and Cu-catalyzed oxidation of aromatic sulfides with H_2O_2 in water [109]. The molybdenum-based catalysts showed better results (yields of 28 to 99%, and *ee*s of 35 to 65%). In a few other accounts on copper-catalyzed asymmetric sulfoxidation, low *ee*s were reported [110–112].

Unlike their behaviors in asymmetric epoxidations, manganese catalysts are poorly represented in asymmetric sulfoxidations with H_2O_2. A few examples include Jacobsen's manganese–salen catalysts of type **43** that catalyzed the oxidation of several alkyl aryl sulfides with high yields and moderate enantioselectivities (34 to 68% *ee*) [113]. Fontecave's manganese-Schiff base complexes **44** that catalyzed the oxidation of aryl methyl sulfides with 20 to 62% *ee* (albeit performing only 1 to 7 catalytic turnovers) [114], and Sun's ternary immobilized manganese–salen system (with sulfide oxidation enantioselectivities of up to 92% *ee*) [115]. The latter system required a 10-fold excess of H_2O_2. Simonneaux's group reported a manganese–porphyrin complex **36b** that catalyzed the oxidation of several prochiral sulfides to sulfoxides with up to 57% *ee* in methanol [116]; the enantioselectivities demonstrated by **36b** were inferior to those of the iron counterpart **36a** [99].

Aerobic asymmetric sulfoxidations promoted by transition metal catalysts are very rare. In 1995, Mukaiyama reported that β-oxo aldiminato–manganese(III) complexes of type **45** (also capable of asymmetric epoxidation of olefins; see Section 2.1; Chapter 2) catalyzed the asymmetric oxidation of sulfides with molecular oxygen in the presence of co-reductant pivalaldehyde [117,118]. Using 12 to 18 mol% of the catalyst, aryl methyl sulfoxides up to 72% *ee* were obtained with moderate to good yields (44 to 94%).

Much later, Katsuki with co-workers found that ruthenium–salen complexes of type **46** are capable of aerobic photopromoted enantioselective oxidation of sulfides

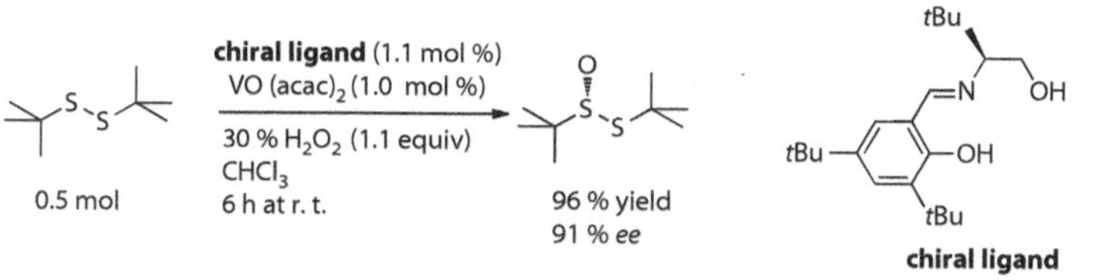

Synthesis of (*S*)-*t*-butyl-*t*-butanethiosulfinate [44].

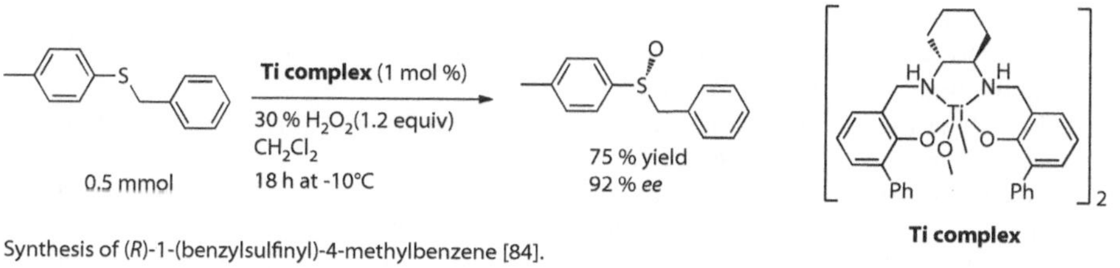

Synthesis of (*R*)-1-(benzylsulfinyl)-4-methylbenzene [84].

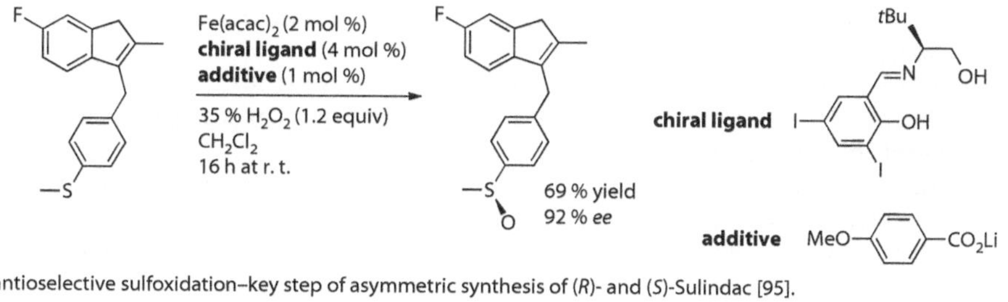

Enantioselective sulfoxidation–key step of asymmetric synthesis of (*R*)- and (*S*)-Sulindac [95].

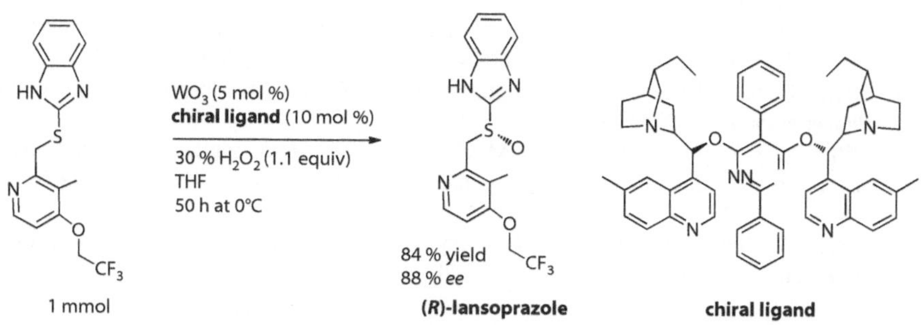

Enantioselective synthesis of anti-ulcer drug (*R*)-lansoprazole from the corresponding prochiral sulfide [104].

FIGURE 3.7 Examples of preparative scale catalytic asymmetric sulfoxidations.

to sulfoxides upon light irradiation without any sacrificial reductant [119]. With 5 mol% catalyst load, the system (with catalyst **46b**) oxidized aryl methyl sulfides and 1,3-dithianes by air oxygen, producing sulfoxides at 18 to 98% yield and 84 to 98% *ee* within 48 hr.

To date, asymmetric sulfoxidation catalysts based on metals other than V, Ti, and Fe and using H_2O_2 or dioxygen have been underrepresented. The results (except for those from Katsuki's Al–salalen systems) are inferior to those from vanadium- and titanium-based catalyst systems based on enantioselectivity, catalyst efficiency, and substrate scope.

REFERENCES

1. Posner, G. H. 1987. Asymmetric synthesis of carbon–carbon bonds using sulfinyl cyclo-alkenones, alkenolides, and pyrones. *Acc. Chem. Res.* 20: 72–78.
2. Patai, S., Rappoport, Z., and Stirling, C. J. M., eds. 1988. *Sulphones and Sulphoxides.* Chichester: John Wiley & Sons.
3. Carreño, M. C. 1995. Applications of sulfoxides to asymmetric synthesis of biologically active compounds. *Chem. Rev.* 95: 1717–1760.
4. Prilezhaeva, E. N. 2000. Sulfones and sulfoxides in the total synthesis of biologically active natural compounds. *Russ. Chem. Rev.* 69: 367–408.
5. Nenajdenko, G., Krasovskiy A. L., and Balenkova, E. S. 2007. The chemistry of sulfi-nyls and sulfonylenones. *Tetrahedron* 63: 12481–12539.
6. Carreño, M. C., Hernández-Torres, G., Ribagorda, M. et al. 2009. Enantiopure sulfox-ides: recent applications in asymmetric synthesis. *Chem. Commun.* 6129–6144.
7. Spencer, C. M. and Faulds, D. 2000. Esomeprazole. *Drugs* 60: 321–329.
8. Cotton, H., Elebring, T., Larsson, M. et al. 2000. Asymmetric synthesis of esomepra-zole. *Tetrahedron Asymmetry* 11: 3819–3825.
9. Rouhi, A. M. 2003. Chirality at work. *Chem. Eng. News* 81: 56.
10. Federsel, H. J. 2003. Facing *chirality* in the 21st century: approaching the challenges in the pharmaceutical industry. *Chirality* 15: S128–S142.
11. Vakily, M., Zhang, W. J., Wu, J. T. et al. 2009. Pharmacokinetics and pharmacodynamics of a known active PPI with a novel dual delayed release technology. Dexlansoprazole MR: a combined analysis of randomized controlled clinical trials. *Curr. Med. Res. Opin.* 25: 627–638.
12. Raju, M. N., Kumar, N. U., Reddy, B. S. et al. 2011. An efficient synthesis of dexlanso-prazole employing asymmetric oxidation strategy. *Tetrahedron Lett.* 52: 5464–5466.
13. Girish, D., Khile, A., Pradhan, N. et al. 2009. Process for the preparation of modafinil enantiomers. Patent Appl. WO 2009/024863 A2, priority date June 26, 2007.
14. Bogan, R. K. 2010. Armodafinil in the treatment of excessive sleepiness. *Expert Opin. Pharmacother.* 11: 993–1002.
15. Miller, M. J. S., Bednar, M. M., and McGiff, J. C. 1983. Renal metabolism of Sulindac, a novel non-steroidal anti-inflammatory agent. *Adv. Prostaglandin Thromboxane Leukot. Res.* 11: 487–491.
16. Legros, J., Dehli, J. R., and Bolm, C. 2005. Applications of catalytic asymmetric sulfide oxidations to the syntheses of biologically active sulfoxides. *Adv. Synth. Catal.* 347: 19–31.
17. Fernandez, I. and Khiar, N. 2003. Recent developments in the synthesis and utilization of chiral sulfoxides. *Chem. Rev.* 103: 3651–3705.
18. Wojaczyńska, E. and Wojaczyński, J. 2010. Enantioselective synthesis of sulfoxides: 2000–2009. *Chem. Rev.* 110: 4303–4356.

19. O'Mahony, G. E., Kelly, P., Lawrence, S. E. et al. 2011. Synthesis of enantio-enriched sulfoxides. *ARKIVOC*, 1–110.
20. Bryliakov, K. P. and Talsi, E. P. 2012. Transition metal-catalyzed asymmetric oxidation of sulfides: from discovery to recent trends. *Curr. Org. Chem.* 16: 1215–1242.
21. Farina, V., Reeves, J. T., Senanayake, C. H. et al. 2006. Asymmetric synthesis of active pharmaceutical ingredients. *Chem. Rev.* 106: 2734–2793.
22. Pitchen, P. and Kagan, H. B. 1984. An efficient asymmetric oxidation of sulfides to sulfoxides. *Tetrahedron Lett.* 25: 1049–1052.
23. Pitchen, P., Desmukh, M., Dunach, E. et al. 1984. An efficient asymmetric oxidation of sulfides to sulfoxides. *J. Am. Chem. Soc.* 106: 8188–8193.
24. Di Furia, F., Modena, G., and Seraglia, R. 1984. Synthesis of chiral sulfoxides by metal-catalyzed oxidation with *t*-butyl hydroperoxide. *Synthesis* 325–326.
25. Katsuki, T. and Sharpless, K. B. 1980. The first practical method for asymmetric epoxidation. *J. Am. Chem. Soc.* 102: 5974–5976.
26. Imagawa, K., Nagata, T., Yamada, T. et al. 1995. Asymmetric oxidation of sulfides with molecular oxygen catalyzed by β-oxo aldiminato–manganese(III) complexes. *Chem. Lett.* 335–336.
27. Nagata, T., Imagawa, K., Yamada, T. et al. 1995. Enantioselective aerobic oxidation of sulfides catalyzed by optically active β-oxo aldiminato–manganese(III) complexes. *Bull. Chem. Soc. Jpn.* 68: 3241–3246.
28. Tanaka, H., Nishikawa, H., Uchida, T. et al. 2010. Photopromoted Ru-catalyzed asymmetric aerobic sulfide oxidation and epoxidation using water as a proton transfer mediator. *J. Am. Chem. Soc.* 132: 12034–12041.
29. Bolm, C. and Bienewald, F. 1995. Asymmetric sulfide oxidation with vanadium catalysts and H_2O_2. *Angew. Chem. Int. Ed. Engl.* 34: 2640–2642.
30. Bolm, C., Schlingloff, G., and Bienewald, F. 1997. Copper- and vanadium-catalyzed asymmetric oxidations. *J. Mol. Catal. A Chem.* 117: 347–350.
31. Bolm, C. and Bienewald, F. 1998. Asymmetric oxidation of dithioacetals and dithioketals catalyzed by a chiral vanadium complex. *Synlett.* 1327–1328.
32. Karpyshev, N. N., Yakovleva, O. D., Talsi, E.P. et al. 2000. Effect of portion-wise addition of oxidant in asymmetric vanadium-catalyzed sulfide oxidation. *J. Mol. Catal. A Chem.* 157: 91–95.
33. Bryliakov, K. P., Karpyshev, N. N., Fominsky, S. A. et al. 2001. ^{51}V and ^{13}C NMR spectroscopic study of the peroxovanadium intermediates in vanadium-catalyzed enantioselective oxidation of sulfides. *J. Mol. Catal. A Chem.*171: 73–80.
34. Blum, S. A., Bergman, R. G., and Ellman, J. A. 2003. Enantioselective oxidation of di-*tert*-butyl disulfide with a vanadium catalyst: progress toward mechanism elucidation. *J. Org. Chem.* 68: 150–155.
35. Ando, R., Nagai, M., Yagyu, T. et al. 2003. Composition and geometry of oxovanadium(IV) and (V)–aminoethanol–Schiff base complexes and stability of their peroxo complexes in solution. *Inorg. Chim. Acta* 351: 107–113.
36. Bryliakov, K. P. and Talsi, E. P. 2003. Intermediates of asymmetric oxidation processes catalyzed by vanadium(V) complexes. *Kinet. Catal.* 44: 334–346.
37. Zeng, Q. L., Wang, H. Q., Weng, W. et al. 2005. Substituent effects and mechanism elucidation of enantioselective sulfoxidation catalyzed by vanadium–Schiff base complexes. *New J. Chem.* 29: 1125–1127.
38. Jeong, Y. C., Kang, E. J., and Ahn, K. H. 2009. Electronic effects of substituents in sulfides: mechanism elucidation of vanadium-catalyzed sulfoxidation. *Bull. Korean Chem. Soc.* 30: 2795–2798.
39. Conte, V., Coletti, A., Floris, B. et al. 2011. Mechanistic aspects of vanadium-catalysed oxidations with peroxides. *Coord. Chem. Rev.* 255: 2165–2177.

40. Skarzewski, J., Ostrycharz, E., and Siedlecka, R. 1999. Vanadium-catalyzed enantiose-lective oxidation of sulfides: easy transformation of *bis*(arylthio)alkanes into C$_2$ sym-metric chiral sulfoxides. *Tetrahedron Asymmetry* 10: 3457–3461.

41. Skarzewski, J., Ostrycharz, E., Siedlecka, R. et al. 2001. Substituted *N*-salicylidene β-amino alcohols: preparation and use as chiral ligands in enantioselective sulfoxida-tion and conjugate addition. *J. Chem. Res. (S),* 263–264.

42. Vetter, A. H. and Berkessel, A. 1998. Schiff-base ligands carrying two elements of chi-rality: matched and mismatched effects in the vanadium-catalyzed sulfoxidation of thio-ethers with hydrogen peroxide. *Tetrahedron Lett.* 39: 1741–1744.

43. Ohta, C., Shimizu, H., Kondo, A. et al. 2002. Vanadium-catalyzed enantioselective sulf-oxidation of methyl aryl sulfides with hydrogen peroxide as terminal oxidant. *Synlett.* 161–163.

44. Liu, G., Cogan, D. A., and Ellman, J. A. 1997. Catalytic asymmetric synthesis of *tert*-butane sulfinamide: application to the asymmetric synthesis of amines. *J. Am. Chem. Soc.* 119: 9913–9914.

45. Gama, Á., Flores-López, L. Z., Aguirre, G. et al. 2003. Oxidation of sulfides to chiral sulfoxides using Schiff base–vanadium(IV) complexes. *ARKIVOC* 11: 4–15.

46. Flores-López, L. Z., Iglesias, A. L., Gama, Á. et al. 2007. Lewis acid–Lewis base bifunc-tional Schiff base titanium and vanadium catalysis for enantioselective cyanosilylation of benzaldehyde and oxidation of sulfides to chiral sulfoxides. *J. Mex. Chem. Soc.* 51: 175–180.

47. Pelotier, B., Anson, M., Campbell, I. B. et al. 2002. Enantioselective sulfide oxidation with H$_2$O$_2$: a solid phase and array approach for the optimisation of chiral Schiff base–vanadium catalysts. *Synlett.* 1055–1060.

48. Baltork, I. M. B., Hill, M., Caggiano, L. et al. 2006. Oxidative kinetic resolution of alkyl aryl sulfoxides. *Synlett.* 3540–3544.

49. Drago, C., Caggiano, L., and Jackson, R. F. W. 2005. Vanadium-catalyzed sulfur oxida-tion and kinetic resolution in the synthesis of enantiomerically pure alkyl aryl sulfox-ides. *Angew. Chem. Int. Ed.* 44: 7221–7223.

50. Zeng, Q., Wang, H., Wang, T. et al. Vanadium-catalyzed enantioselective sulfoxidation and concomitant highly efficient kinetic resolution provide high enantioselectivity and acceptable yields of sulfoxides. *Adv. Synth. Catal.* 347: 1933–1936.

51. Kelly, P., Lawrence, S. E., and Maguire, A. R. 2006. Kinetic resolution in vanadium-catalyzed sulfur oxidation as an efficient route to enantiopure aryl benzyl sulfoxides. *Synlett.* 1569–1573.

52. Kelly, P., Lawrence, S. E., and Maguire, A. R. 2006. Asymmetric synthesis of aryl benzyl sulfoxides by vanadium-catalysed oxidation: a combination of enantioselective sulfide oxidation and kinetic resolution in sulfoxide oxidation. *Eur. J. Org. Chem.* 4500–4509.

53. Zeng, Q., Gao, Y., Dong, J. et al. 2011. Vanadium-catalyzed enantioselective oxidation of allyl sulfides. *Tetrahedron Asymmetry* 22: 717–721.

54. Liu, H., Wang, M., Wang, Y. et al. 2008. Influence of substituents in the salicylaldehyde-derived Schiff bases on vanadium-catalyzed asymmetric oxidation of sulfides. *Appl. Organomet. Chem.* 22: 253–257.

55. Gao, A., Wang, M., Wang, D. et al. 2006. Asymmetric oxidation of sulfides catalyzed by vanadium(IV) complexes of dibromo- and diiodo-functionalized chiral Schiff bases. *Chinese J. Catal.* 27: 743–748.

56. Liu, H., Wang, M., Wang, Y. et al. 2009. Asymmetric oxidation of sulfides with hydro-gen peroxide catalyzed by a vanadium complex of a new chiral NOO ligand. *Catal. Commun.* 11: 294–297.

57. Hsieh, S. H., Kuo, Y. P., and Gau, H. M. 2007. Synthesis, characterization, and structures of oxovanadium(V) complexes of Schiff bases of β-amino alcohols as tunable catalysts for the asymmetric oxidation of organic sulfides and asymmetric alkynylation of aldehydes. *Dalton Trans.* 97–106.

58. Suresh, P., Srimurugan, S., Babu, B. et al. 2007. Asymmetric sulfoxidation of prochiral sulfides using amino alcohol-derived chiral C_3-symmetric trinuclear vanadium Schiff base complexes. *Tetrahedron Asymmetry* 18: 2820–2827.

59. Wang, Y., Wang, M., Wang, Y. et al. 2010. Highly enantioselective sulfoxidation with vanadium catalysts of Schiff bases derived from bromo- and iodo-functionalized hydroxynaphthaldehydes. *J. Catal.* 273: 177–181.

60. Wu, Y., Liu, J., Li, X. et al. 2009. Vanadium-catalyzed asymmetric oxidation of sulfides using Schiff base ligands derived from β-amino alcohols with two stereogenic centers. *Eur. J. Org. Chem.* 2607–2610.

61. Wu, Y., Mao, F., Meng, F. et al. 2011. Enantioselective vanadium-catalyzed oxidation of 1,3-dithianes from aldehydes and ketones using β-amino alcohol-derived Schiff base ligands. *Adv. Synth. Catal.* 353: 1707–1712.

62. Jeong, Y. C., Choi, S., Hwang, Y. D. et al. 2004. Enantioselective oxidation of sulfides with hydrogen peroxide catalyzed by vanadium complex of sterically hindered chiral Schiff bases. *Tetrahedron Lett.* 45: 9249–9252.

63. Jeong, Y. C., Huang, Y. D., Choi, S. et al. 2005. Synthesis of sterically controlled chiral β-amino alcohols and their application to the catalytic asymmetric sulfoxidation of sulfides. *Tetrahedron Asymmetry* 16: 3497–3501.

64. Khiar, N., Mallouk, S., Valdivia, V. et al. 2007. Enantioselective organocatalytic oxidation of functionalized sterically hindered disalfides, *Org. Lett.* 9: 1255–1258.

65. Cucciolito, M. E., Del Litto, R., Roviello, G. et al. 2005. O,N,O′ tridentate ligands derived from carbohydrates in the V(IV)-promoted asymmetric oxidation of thioanisole. *J. Mol. Catal. A Chem.* 236: 176–181.

66. Lippold, I., Becher, J., Klemm, D. et al. 2009. Chiral oxovanadium(V) complexes with a 6-amino-6-deoxyglucopyranoside-based Schiff base ligand: catalytic asymmetric sulfoxidation and structural characterization. *J. Mol. Catal. A Chem.* 299: 12–17.

67. Koneva, E. A., Volcho, K. P., Korchagina, D. V. et al. 2008. New chiral Schiff bases derived from (+)- and (−)-α-pinenes in the metal complex catalyzed asymmetric oxidation of sulfides. *Russ. Chem. Bull.* 57: 108–117.

68. Koneva, E. A., Volcho, K. P., Korchagina, D. V. et al. 2009. Synthesis of new chiral Schiff bases from (+)-3-carene and their use in asymmetric oxidation of sulfides catalyzed by metal complexes. *Russ. J. Org. Chem.* 45: 815–824.

69. Green, S. D., Monti, C., Jackson, R. F. W. et al. 2001. Discovery of new solid phase sulfur oxidation catalysts using library screening. *Chem. Commun.* 2594–2595.

70. Barbarini, A., Maggi, R., Muratori, M. et al. 2004. Enantioselective sulfoxidation catalyzed by polymer-supported chiral Schiff base–VO(acac)$_2$ complexes. *Tetrahedron Asymmetry* 15: 2467–2473.

71. Bryliakov, K. P. and Talsi, E. P. 2008. Titanium–salan-catalyzed asymmetric oxidation of sulfides and kinetic resolution of sulfoxides with H$_2$O$_2$ as oxidant. *Eur. J. Org. Chem.* 3369–3376.

72. Adão, P., Pessoa, J. C., Henriques, R. T. et al. 2009. Synthesis, characterization, and application of vanadium–salan complexes in oxygen transfer reactions. *Inorg. Chem.* 48: 3542–3561.

73. Volcho, K. P. and Salakhutdinov, N. F. 2009. Asymmetric oxidation of sulfides catalyzed by titanium and vanadium complexes in the synthesis of biologically active sulfoxides. *Russ. Chem. Rev.* 78: 457–464.

74. Stingl, K. A., Weiss, K. M., and Tsogoeva, S. B. 2012. Asymmetric vanadium- and iron-catalyzed oxidations: new mild (*R*)-modafinil synthesis and formation of epoxides using aqueous H_2O_2 as a terminal oxidant. *Tetrahedron* 68: 8493–8501.
75. Colombo, A., Marturano, G., Pasini, A. 1986. Chiral induction in the oxidation of thioanisole with chiral oxotitanium(IV)–Schiff-base complexes as catalysts: the importance of the conformation of the ligands. *Gazz. Chim. Ital.* 116: 35–40.
76. Saito, B. and Katsuki, T. 2001. Ti(salen)-catalyzed enantioselective sulfoxidation using hydrogen peroxide as a terminal oxidant. *Tetrahedron Lett.* 42: 3873–3876.
77. Saito, B. and Katsuki, T. 2001. Mechanistic consideration of Ti(salen)-catalyzed asymmetric sulfoxidation. *Tetrahedron Lett.* 42: 8333–8336.
78. Tanaka, T., Saito, B., and Katsuki, T. 2002. Highly enantioselective oxidation of cyclic dithioacetals by using a Ti(salen) and urea·hydrogen peroxide system. *Tetrahedron Lett.* 43: 3259–3262.
79. Bryliakov, K. P. and Talsi, E. P. 2007. Asymmetric oxidation of sulfides with H_2O_2 catalyzed by titanium complexes with aminoalcohol derived Schiff bases. *J. Mol. Catal. A Chem.* 264: 280–287.
80. Bryliakov, K. P., Nuzhdin, A. L., and Talsi, E. P. 2007. Fe(III), Ti(IV), and Zr(IV)-catalyzed asymmetric oxidations of thioethers and product separation via enantioselective sorption on chiral metal-organic frameworks. In *Catalysis: Fundamentals and Applications: Abstracts of the III International Conference.* Novosibirsk: Boreskov Institute of Catalysis, pp. 155–156.
81. Wang, Y., Wang, M., Wang, L. et al. 2011. Asymmetric oxidation of sulfides with H_2O_2 catalyzed by titanium complexes of Schiff bases bearing a dicumenyl salicylidenyl unit. *Appl. Organomet. Chem.* 25: 325–330.
82. Bera, P. K., Ghosh, D., Abdi, S. H. R. et al. 2012. Titanium complexes of chiral amino alcohol-derived Schiff bases as efficient catalysts in asymmetric oxidation of prochiral sulfides with hydrogen peroxide as an oxidant. *J. Mol. Catal. A Chem.* 361: 36–44.
83. Abdi, S. H. R. et al. reported use of 0.75 to 1.75 *M* concentrations of H_2O_2 in CH_2Cl_2, whereas such high concentrations could not be achieved due to limited solubility of highly polar H_2O_2 in non-polar CH_2Cl_2; in practice, sulfoxidation in CH_2Cl_2 means sulfoxidation in a two-phase system.
84. Bryliakov, K. P. and Talsi, E. P. 2011. Catalytic enantioselective oxidation of bulky alkyl aryl thioethers with H_2O_2 over titanium–salan catalysts. *Eur. J. Org. Chem.* 4693–4698.
85. Talsi, E. P. and Bryliakov, K. P. 2013. Titanium–salan-catalyzed asymmetric sulfoxidations with H_2O_2: design of more versatile catalysts. *Appl. Organomet. Chem.* 27: 239–244.
86. Adão, P., Avecilla, F., Bonchio, M. et al. 2010. Titanium(IV)–salan catalysts for asymmetric sulfoxidation with hydrogen peroxide. *Eur J. Inorg. Chem.* 5568–5578.
87. Barman, S., Patil, S., and Levy, C. J. 2012. Asymmetric sulfoxidation of thioanisole by helical Ti(IV)–salan catalysts. *Chem. Lett.* 41: 974–975.
88. Panda, M. K., Shaikh, M. M., and Ghosh, P. 2010. Controlled oxidation of organic sulfides to sulfoxides under ambient conditions by a series of titanium isopropoxide complexes using environmentally benign H_2O_2 as an oxidant. *Dalton Trans.* 39: 2428–2440.
89. Duboc-Toia, C., Menage, S., Lambeaux, C. et al. 1997. μ-Oxo diferric complexes as oxidation catalysts with hydrogen peroxide and their potential in asymmetric oxidation. *Tetrahedron Lett.* 38: 3727–3730.
90. Duboc-Toia, C., Menage, S., Ho, R. Y. H. et al. 1999. Enantioselective sulfoxidation as a probe for a metal-based mechanism in H_2O_2-dependent oxidations catalyzed by a di-iron complex. *Inorg. Chem.* 38: 1261–1268.
91. Mekmouche, Y., Hummel, H., Ho, R. Y. N. et al. 2002. Sulfide oxidation by hydrogen peroxide catalyzed by iron complexes: two metal centers are better than one. *Chem. Eur. J.* 8: 1196–1204.

92. Legros, J. and Bolm, C. 2003. Iron-catalyzed asymmetric sulfide oxidation with aqueous hydrogen peroxide. *Angew. Chem. Int. Ed. Engl.* 115: 5487–5489.
93. Legros, J. and Bolm, C. 2004. Highly enantioselective iron-catalyzed sulfide oxidation with aqueous hydrogen peroxide under simple reaction conditions. *Angew. Chem. Int. Ed. Engl.* 116: 4225–4228.
94. Legros, J. and Bolm, C. 2005. Investigations of iron-catalyzed asymmetric sulfide oxidation. *Chem. Eur. J.* 11: 1086–1092.
95. Korte, A., Legros, J., and Bolm, C. 2004. Asymmetric synthesis of Sulindac by iron-catalyzed sulfoxidation. *Synlett* 13: 2397–2399.
96. Egami, H. and Katsuki, T. 2007. Fe(salan)-catalyzed asymmetric oxidation of sulfides with hydrogen peroxide in water. *J. Am. Chem. Soc.* 129: 8940–8941.
97. Egami, H. and Katsuki, T. 2008. Optimization of asymmetric oxidation of sulfides with the Fe(salan) complex in water and the expanded scope of its application. *Synlett* 1543–1546.
98. Li, B., Bai, S. Y., Wang, P. et al. 2011. Encapsulation of chiral Fe(salan) in nanocages with different microenvironments for asymmetric sulfide oxidation. *Phys. Chem. Chem. Phys.* 13: 2504–2511.
99. Le Maux, P. and Simonneaux, G. 2011. First enantioselective iron–porphyrin-catalyzed sulfide oxidation with aqueous hydrogen. *Chem. Commun.* 47: 6957–6959.
100. Yamaguchi, T., Matsumoto, K., Saito, B. et al. 2007. Asymmetric oxidation catalysis by a chiral Al(salalen) complex: highly enantioselective oxidation of sulfides with aqueous hydrogen peroxide. *Angew. Chem. Int. Ed.* 46: 4729–4731.
101. Matsumoto, K., Yamaguchi, T., and Katsuki, T. 2008. Asymmetric oxidation of sulfides under solvent-free or highly concentrated conditions. *Chem. Commun.* 1704–1706.
102. Matsumoto, K., Yamaguchi, T., Fujisaki, J. et al. 2008. Aluminum oxidation catalysis under aqueous conditions: highly enantioselective sulfur oxidation catalyzed by Al(salalen) complexes. *Chem. Asian J.* 3: 351–358.
103. Fujisaki, J., Matsumoto, K., Matsumoto, K. et al. 2011. Catalytic asymmetric oxidation of cyclic dithioacetals: highly diastereo- and enantioselective synthesis of the *S*-oxides by a chiral aluminum(salalen) complex. *J. Am. Chem. Soc.* 133: 56–61.
104. Thakur V. V. and Sudalai, A. 2003. WO$_3$–30% H$_2$O$_2$–cinchona alkaloids: a new heterogeneous catalytic system for the asymmetric oxidation of sulfides and the kinetic resolution of racemic sulfoxides. *Tetrahedron Asymmetry* 14: 407–410.
105. Zhang, Y., Sun, J. T., and Zhu, C. J. 2006. WO$_3$–salan–30% H$_2$O$_2$: a heterogeneous catalytic system for the asymmetric oxidation of sulfides. *Chin. Chem. Lett.* 17: 1173–1176.
106. Bigi, F., Gunaratne, H. Q. N., Quarantelli, C. et al. 2011. Chiral ionic liquids for catalytic enantioselective sulfide oxidation. *C. R. Chimie* 14: 685–687.
107. Kelly, P., Lawrence, S. E., Maguire, A. R. 2007. Chemoselectivity and enantioselectivity in copper-catalysed oxidation of aryl benzyl sulfides. *Synlett* 10:1501–1506.
108. O'Mahony, G., Ford, A., and Maguire, A. 2012. Copper-catalyzed asymmetric oxidation of sulfides. *J. Org. Chem.* 77: 3288–3296.
109. Sakuraba, H. and Maekawa, H. 2006. Enantioselective oxidation of sulfides catalyzed by chiral Mo^V and Cu^{II} complexes of catechol-appended β-cyclodextrin derivatives in water. *J. Inclusion Phenom. Macrocyclic Chem.* 54: 41–45.
110. Zhu, H. B., Dai, Z. Y., Huang, W. et al. 2004. Chiral copper(II) complexes of optically active Schiff bases: syntheses, crystal structure and asymmetric oxidation of methyl phenyl sulfide with hydrogen peroxide. *Polyhedron* 23: 1131–1137.
111. Ayala, V., Corma, A., Iglesias, M. et al. 2004. Mesoporous MCM41-heterogenised (salen)Mn and Cu complexes as effective catalysts for oxidation of sulfides to sulfoxides: isolation of a stable supported Mn(V)=O complex responsible for the catalytic activity. *J. Mol. Catal. A Chem.* 221: 201–208.

112. Punniyamurthy, T. and Rout, L. 2008. Recent advances in copper-catalyzed oxidation of organic compounds. *Coord. Chem. Rev.* 252: 134–154.

113. Palucki, M., Hanson, P., and Jacobsen, E. N. 1992. Asymmetric oxidation of sulfides with H_2O_2 catalyzed by (salen)Mn(III) complexes. *Tetrahedron Lett.* 33: 7111–7114.

114. Schoumacker, S., Hamelin, O., Pacaut, J. et al. 2003. Catalytic asymmetric sulfoxidation by chiral manganese complexes: acetylacetonate anions as chirality switches. *Inorg. Chem.* 42: 8110–8116.

115. Zhang, Z., Guan, F., Huang, X. et al. 2012. New ternary immobilization of chiral sulfonato-(salen)manganese(III) complex for aqueous asymmetric oxidation reactions. *J. Mol. Catal. A Chem.* 363: 343–353.

116. Srour, H., Jalkh, J., Le Maux, P. et al. G. 2013. Asymmetric oxidation of sulfides by hydrogen peroxide catalyzed by chiral manganese porphyrins in water and methanol solution. *J. Mol. Catal. A Chem.* 370: 75–79.

117. Imagawa, K., Nagata, T., Yamada, T. et al. 1995. Asymmetric oxidation of sulfides with molecular oxygen catalyzed by β-oxo aldiminato manganese(III) complexes. *Chem. Lett.* 335–336.

118. Nagata, T., Imagawa, K., Yamada, T. et al. 1995. Enantioselective aerobic oxidation of sulfides catalyzed by optically active β-oxo aldiminato manganese(III) complexes. *Bull. Chem. Soc. Jpn.* 68: 3241–3246.

119. Tanaka, H., Nishikawa, H., Uchida, T. et al. 2010. Photopromoted Ru-catalyzed asymmetric aerobic sulfide oxidation and epoxidation using water as a proton transfer mediator. *J. Am. Chem. Soc.* 132: 12034–12041.

4 Miscellaneous Transition Metal-Catalyzed Asymmetric Oxidations

This chapter presents an overview of various transition metal-catalyzed oxidative transformations with H_2O_2 or O_2 used as the terminal oxidant, such as *cis*-dihydroxylations, Baeyer–Villiger oxidations, oxidative kinetic resolution of secondary alcohols, desymmetrization of *meso*-diols, and aerobic coupling of 2-naphthols.

CIS-DIHYDROXYLATIONS OF OLEFINS

Diols are common motifs in many natural products and intermediates [1]. The catalytic asymmetric dihydroxylation of olefins to 1,2-diols emerged as an important general method for olefin functionalization. This approach has been used to construct synthetically valuable building blocks and key intermediates in natural product synthesis [2,3]. Most synthetic applications rely on the direct osmium-catalyzed *cis*-dihydroxylation developed by Sharpless and co-workers [2–5]. The main disadvantages of osmium-based systems are their high toxicities and the volatility of osmium tetroxide. Catalyst systems based on complexes of other metals continue to be explored.

Among non-osmium-mediated *cis*-dihydroxylations with H_2O_2, several complexes of iron and manganese are known. The iron catalysts appeared after extensive research directed at modeling the reactivities of Rieske dioxygenases (affording *cis*-1,2-diols) by synthetic non-heme iron complexes. Most known iron-based catalyst systems afford racemic diols [1]; however, Que and co-workers reported that iron (II) triflate complex **1** (Figure 4.1) catalyzed the *cis*-dihydroxylation of various alkenes with 10 to 20 equivalents of H_2O_2 (Table 4.1); the *ee* levels ranged from 3 to 82% [6]. The diol yields were low to moderate, approaching 11.2 mmol of diol per mmol of catalyst formed (upon loading 0.7 mmol of Fe complex, 700 mmol of olefin and 7 to 14 mmol of H_2O_2).

The authors noticed that *cis*-β-topology of complex 1 was essential for achieving high diol (versus epoxide) selectivity [6]. However, this requirement is not peremptory. Later studies showed that iron (II) triflate complexes with chiral ligands **2** and **3** featuring *cis*-α ligand topology also yielded olefin oxidation with H_2O_2 in favor of *cis*-diol formation, with up to 97% *ee* [7]. The authors ascribed the improved enantioselection of [(**3**)Fe(OTf)$_2$] to the more rigid bipyrrolidine ligand backbone (compared to the 1,2-diaminocyclohexane ligand in **1**) and to the *cis*-α-topology of [(**3**)Fe(OTf)$_2$] versus the *cis*-β topology of **1**.

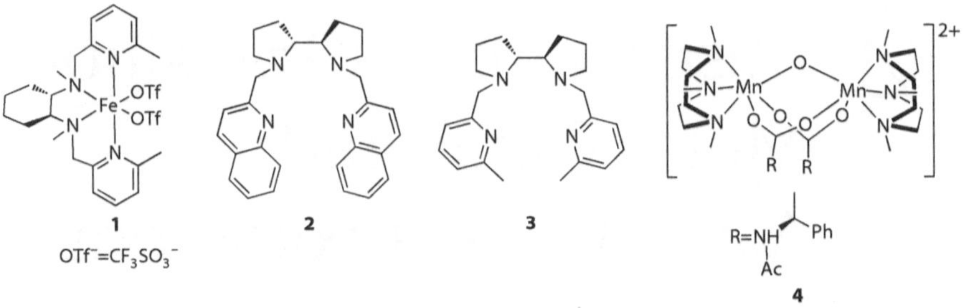

FIGURE 4.1 Iron and ruthenium catalysts and chiral ligands for enantioselective *cis*-dixydroxylations with H_2O_2.

TABLE 4.1

Asymmetric cis-Dihydroxylations with H_2O_2 Catalyzed by Metal Complexes

N	Substrate	Catalyst	*cis*-Diol[a]	*ee* (%)	Ref.
1	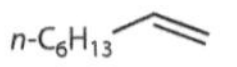 n-C$_6$H$_{13}$	1	8.1	60	[6]
		[(3)Fe(OTf)$_2$]	6.4	76	[7]
2	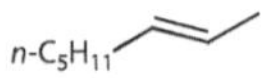 n-C$_5$H$_{11}$	1	7.5	82	[6]
3	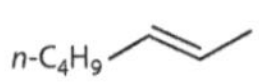 n-C$_4$H$_9$	1	7.5	79	[6]
		[(2)Fe(OTf)$_2$]	3.6	78	[7]
		[(3)Fe(OTf)$_2$]	5.2	97	[7]
4	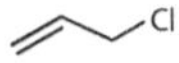 Cl	[(3)Fe(OTf)$_2$]	7.9	70	[7]
5	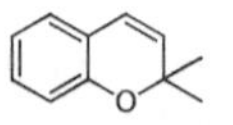	4[a]	[b]	54	[8]

[a] *Cis*-diol yield in mmol of diol/mmol of catalyst.
[b] Conversion of 55% reported.

They also noticed that an increased steric bulk of [(2)Fe(OTf)$_2$] and [(3)Fe(OTf)$_2$] made the complexes more diol-selective than **1** [7]. The only manganese-based asymmetric *cis*-dihydroxylation system was reported by Feringa and co-workers who demonstrated that binuclear manganese complexes of type **4** (0.4 mol% load) with *N*-Boc or *N*-Ac protected chiral α-amino acids (4.0 mol%) catalyzed the asymmetric *cis*-dihydroxylation of 2,2-dimethylchromene with 50% H_2O_2 in low to moderate optical yields (28 to 54% *ee*) [8].

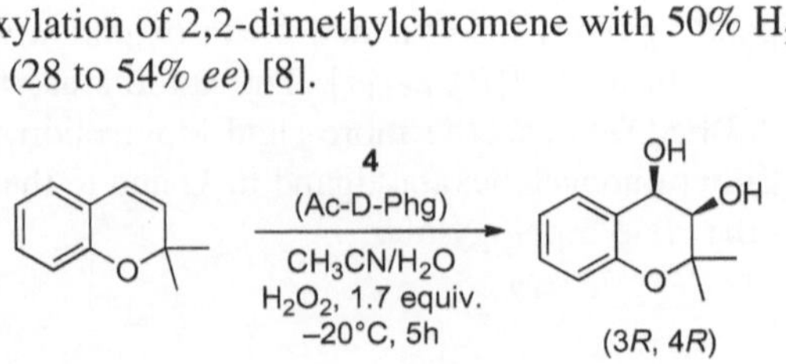

BAEYER–VILLIGER OXIDATIONS

The Baeyer–Villiger oxidation discovered in 1899 [9] is a powerful synthetic method allowing C-C bond breakage with the introduction of an oxygen atom that converts ketones to the corresponding esters or lactones. Initially designed with persulfuric acid as the oxidant, the reaction upgraded several times to come eventually to more sustainable oxidants such as hydrogen peroxide and dioxygen [10,11]. Also, catalytic and asymmetric versions of the Baeyer–Villiger oxidations appeared.

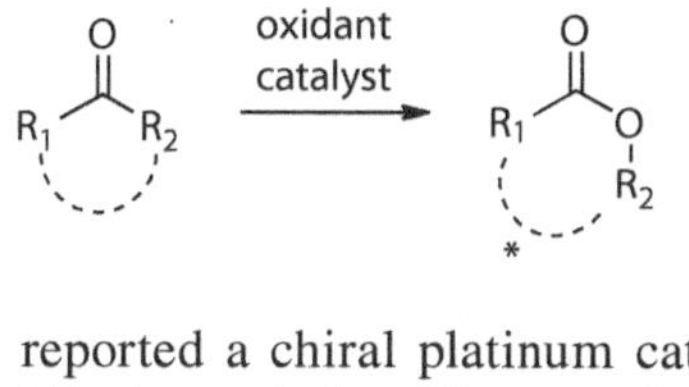

In 1994, Strukul et al. reported a chiral platinum catalyst **5** (Figure 4.2) that was active in the oxidative kinetic resolution of a racemic mixture of cyclic 2-alkyl

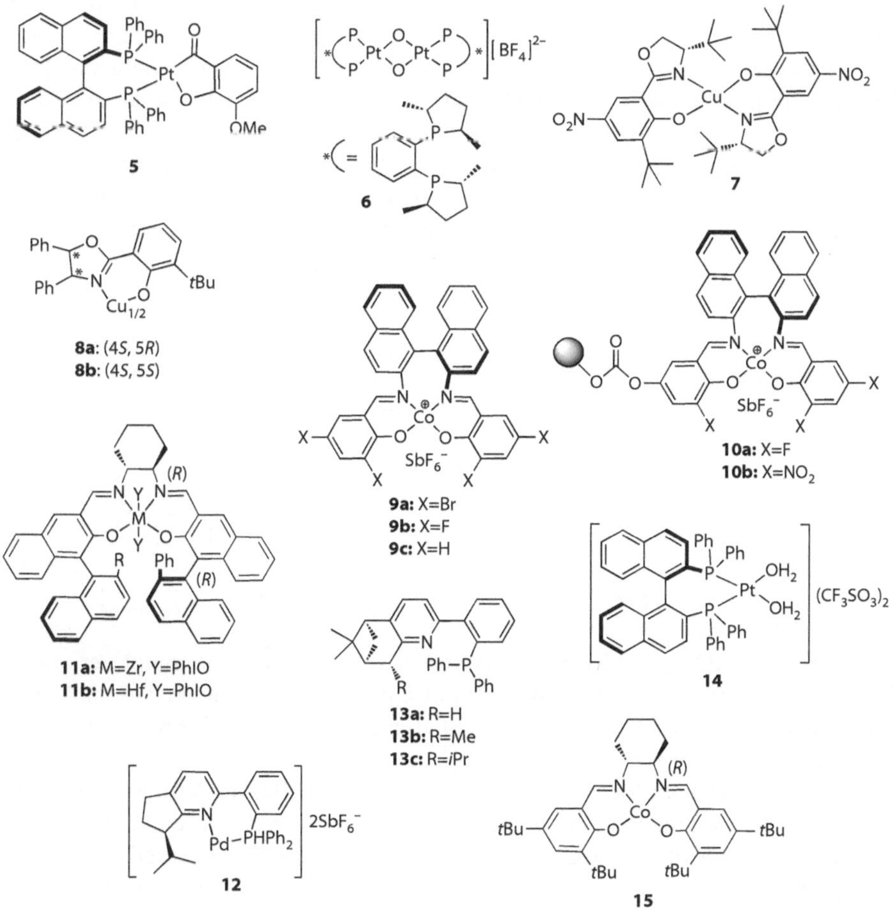

FIGURE 4.2 Catalysts and chiral ligands for enantioselective Baeyer–Villiger oxidations.

substituted cyclopentanones and 2-methylcyclohexanone with aqueous hydrogen peroxide to form chiral lactones with low to moderate enantioselectivities (up to 58% *ee* of the resulting lactone). The reactions were conducted in neat ketone [12]. A series of platinum (II) complexes of type **6** (with various chiral diphosphine moieties) was tested in the catalytic desymmetrization of substituted *meso*-cyclohexanones with H_2O_2 to yield the corresponding lactones with up to 80% *ee* (Table 3.2) [13,14]. Another by-product of the Baeyer–Villiger oxidation is an enantio-enriched ketone (the less reactive enantiomer) that remains mostly unreacted during oxidation.

Bolm and co-workers reported the first copper catalyzed (with complexes of type **7**) asymmetric Baeyer–Villiger oxidation of substituted 2-aryl cyclohexanones with molecular oxygen in the presence of a pivalaldehyde sacrificial reductant [15]. Catalyst **7** (1 mol%) produced chiral lactones with up to 47% yield and up to 69% *ee* (Table 4.2). The same complex **7** catalyzed the oxidation of saturated cyclobutanones to yield a mixture of regioisomeric (normal and abnormal) lactones in comparable amounts (Table 4.2) with high yield and good to high *ees* [16]. The oxidation of 3-substituted cyclobutanones proceeded with moderate enantioselectivities (up to 47% *ee*) [17].

Later, Feng and Jiang with co-workers reported copper oxazoline complexes of type 8 that catalyzed the asymmetric oxidation of 2-phenylcyclohexanone with molecular oxygen in the presence of 1.5 equivalents of aliphatic aldehydes. The reaction demonstrated poor enantioselectivities (below 26% *ee*) and conversions (2 to 62%) [18].

Katsuki and Uchida focused on the enantioselective Baeyer–Villiger oxidations of 3-phenylcyclobutanone in the presence of various catalysts. Several chiral cobalt (II)–salen catalysts of type **9** (5 mol%) were screened, and **9b** appeared as the most enantioselective [19,20]. Oxidations with urea hydroperoxide afforded lactones with generally higher optical purities than with H_2O_2. Ethanol was chosen as the optimal solvent, affording the corresponding lactone in good yields (72 to 92%) and moderate

TABLE 4.2
Asymmetric Baeyer–Villiger Oxidations with H_2O_2 and O_2

N	Substrate	Catalyst	Oxidant	Lactone Yield (%)	ee (%)	Ref.
1		5	H_2O_2	4	58	[12]
2		5	H_2O_2	9	45	[12]
3		6	H_2O_2	9 (6)[a]	80 (11)[a]	[14]
4		7	O_2/PIA	47	69	[15]
5		7	O_2/PIA	53	65	[15]
6		7	O_2/PIA	46/15[b]	67/92[b]	[16]
7		7	O_2/PIA	26/33[b]	59/93	[16]
8		7	O_2/PIA	19/13[b]	76/95	[16]
9		9b	UHP	72	77	[16]
		10b	UHP	67	83	[20]
		11a	UHP	80	85	[21]
		11b	UHP	62	82	[22]
		12	UHP	91	80	[23]
		13c/PdCl$_2$/AgSbF$_6$	UHP	97	81	[24]
		14	H_2O_2	99	56	[25]
10		9b	H_2O_2	75	78	[19]
		10b	H_2O_2	66	79	[20]
		13c/PdCl$_2$/AgSbF$_6$	UHP	91	58	[24]
11		12	UHP	94	83	[23]
		13/PdCl$_2$/AgSbF$_6$	UHP	83	71	[24]
12		14	H_2O_2	8	92	[25]

Note: PIA = pivaldehyde. UHP = urea hydroperoxide.

[a] Yield and *ee* of minor lactone shown in parentheses.

[b] Yield normal/yield abnormal; *ee* normal/*ee* abnormal; see text.

to good *ee* values (50 to 77%) [19,20]. Later, Sandaroos and Goldani reported that polystyrene-supported cobalt (II)–salen complexes of type **10** demonstrated higher enantioselectivities in the oxidations of 3-arylcyclobutanone than their unsupported counterparts [21].

Katsuki and Uchida reported that the second-generation zirconium–salen catalyst **11a** demonstrated substantially higher enantioselectivities than the cobalt catalysts **9**. The former, used in 5 mol%, catalyzed the oxidation of 3-phenylcyclobutanone with up to 87% *ee*; dichloromethane and chlorobenzene were the best solvents [22]. The structurally related hafnium (IV) complex **11b** demonstrated rather similar reactivity toward 3-phenylcyclobutanone [23]. Zirconium–salen complexes also promoted parallel kinetic resolution of racemic bicyclo[4.2.0]octan-7-one, affording normal and abnormal lactones in up to 85% *ee* and >99% *ee*, respectively, and an unreacted ketone with 77 to 94% *ee* [22].

Katsuki and co-workers also reported that cationic palladium (II) complex **12** formed in situ by reacting $PdCl_2(PhCN)$ (5 mol%) and the chiral ligand (5.5 mol%), and $AgSbF_6$ (10 mol%) catalyzed the oxidation of 3-aryl substituted cyclobutanones with UHP, affording the corresponding lactones in 76 to 94% yield and 73 to 83% *ee*. Tricyclic cyclobutanone was oxidized in >99% *ee* [24].

Later, Malkov with co-workers tested terpene-derived ligands in Baeyer–Villiger oxidations according to the same synthetic protocol. They reported that terpene-derived ligands of type **13** could also serve as effective chiral inducers in palladium-catalyzed oxidation of 3-substituted cyclobutanones and produced good to excellent yields (>99%) and moderate to good *ee*s (up to 81%) of corresponding lactones [25].

Strukul and co-workers reported that chiral platinum diphosphine complexes of type **14** catalyzed the enantioselective Baeyer–Villiger oxidation of substituted cyclobutanones and cyclohexanones in water with H_2O_2 [26]. High yields (up to 99%) were achieved only with cyclobutanones, whereas less reactive cyclohexanones demonstrated poor yields (3 to 30%). The catalyst system exhibited a substantial *ee* enhancement when replacing chlorinated organic solvents (dichloroethane, dichloromethane) [27] with aqueous media. For the solubilization of the hydrophobic catalyst and substrate, micelles were used as surfactants and nanoscopic reactors [26]. A similar approach was applied when using the cobalt–salen catalyst **15** in aqueous media [28]; in common organic solvents, the latter catalyst was inefficient. The yields and enantioselectivities in the oxidation of 3-substituted cyclobutanones were poor; moderate to high yields and *ee*s were reported only for the oxidation of bicyclic cyclobutanones like *cis*-bicyclo(3.2.0)hept-2-en-6-one [28].

In recent years, substantial progress was achieved with metal-catalyzed asymmetric Baeyer–Villiger oxidations. However, examples of highly enantioselective and efficient metal-based catalyst systems remain rather rare in the literature; existing systems demonstrate narrow substrate scope (mostly limited to reaction-prone substituted cyclobutanones). To date, no metal complexes are considered as catalysts for practical asymmetric Baeyer–Villiger oxidations with H_2O_2 or O_2. More or less successful alternatives are biocatalytic and organocatalytic [11,29–32] oxidations; the latter types will be discussed in Chapter 5.

OXIDATIVE KINETIC RESOLUTION OF SECONDARY ALCOHOLS AND DESYMMETRIZATION OF MESO-DIOLS

This section covers the oxidations of racemic secondary alcohols in cases in which the oxidations of two enantiomers (performed in the presence of chiral catalysts) occur at different rates, thus yielding a corresponding carbonyl compound along with a residual enantio-enriched alcohol [33]. In addition to this typical case of kinetic resolution of racemic secondary alcohols, catalyst systems for desymmetrization of *meso-* and prochiral diols are overviewed.

Sigman and co-workers reported that in situ-formed palladium(II) complexes with naturally occurring amines [(-)-sparteine **16**; Figure 4.3] catalyzed the oxidative

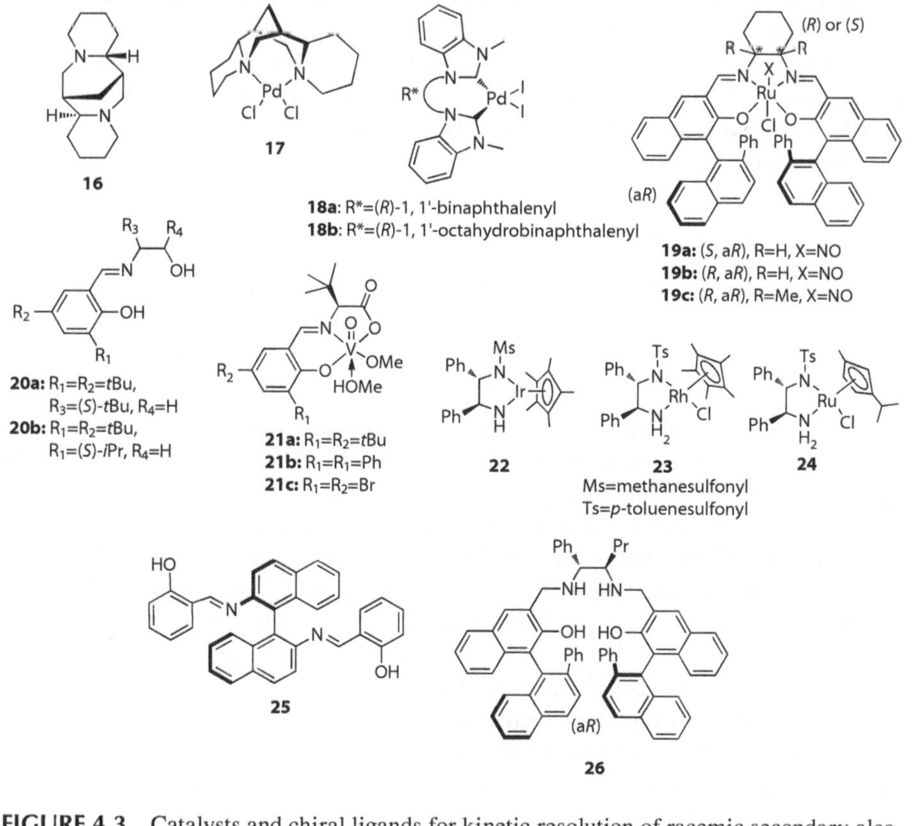

FIGURE 4.3 Catalysts and chiral ligands for kinetic resolution of racemic secondary alcohols and desymmetrization of *meso*-diols.

kinetic resolution of secondary alcohols with molecular oxygen in dichloroethane [34]. A series of benzylic secondary alcohols demonstrated moderate to good k_{rel}, to yield the corresponding ketone and recover the enantiomerically enriched alcohols with 66 to 99% *ee* (Table 4.3).

This catalyst system also performed the oxidative desymmetrization of *meso*-1,3-diphenylpropane-1,3-diol with 82% *ee* and 69% yield. Later, the authors discovered that a palladium(II)–(-)-sparteine complex **17** could promote the asymmetric kinetic resolution only in the presence of exogenous (-)-sparteine, and proposed a dual role of the latter: (1) as a chiral ligand and (2) as a base for β-hydrogen elimination from the palladium–alkoxide intermediate [34,35]. Complex **17**, in the presence of (-)-sparteine, was further used as a catalyst in an extensive study of a broad range of alcohol substrates in the media of *t*-butanol [36].

Simultaneously, palladium–sparteine-catalyzed oxidative kinetic resolution of secondary alcohols was studied by Stolz and Ferreira [37]. They reported a similar catalyst formed in situ from the Pd(II) source and (-)-sparteine that operated in toluene solution in the presence of 3Å molecular sieves. They screened several racemic alcohol mixtures and reported isolated yields of enantio-enriched alcohols of 29 to 49% (i.e., approaching the theoretical yield), with optical purities of 68 to 99% *ee* [37]. The oxidation of 1-naphthalen-2-yl-ethanol was conducted at 5.0 g scale to recover 2.2 g of (-)-1-naphthalen-2-yl-ethanol with 44% yield and 99% *ee* after the first cycle. The ketone was separated, reduced with NaBH$_4$, and again involved in oxidative kinetic resolution. After two cycles, the yield of the optically pure alcohol (99% *ee*) was 68%.

The same group reported that the addition of Cs$_2$CO$_3$ and *t*-butanol resulted in a dramatic acceleration of the reaction [38], so that the latter could be conducted at 60°C (instead of 80°C), typically with comparable selectivity. Further improvements were achieved by using 1 atm ambient air as the oxidant and chloroform as

TABLE 4.3
Oxidative Kinetic Resolution of Racemic Secondary Alcohols

N	Substrate	Catalyst	Additive	Conversion (%)	ee (%)	Ref.
1	OH Ph–Me	Pd(II)/(-)-sparteine	—	66	98	[34]
		Pd(II)/(-)-sparteine	—	60	99	[37]
		18b	Cs$_2$CO$_3$	67	93	[45]
		19b	—	64	95	[47]
		19b	a	61	98	[48]
		22	—	48[b]	98	[56]
		23	—	46[b]	93	[56]
		26	1-naphthol	55	98	[61]
2	OH Ph–Et	Pd(II)/(-)-sparteine	—	58	89	[34]
		Pd(II)/(-)-sparteine	—	59	93	[37]
		Pd(II)/(-)-sparteine	Cs$_2$CO$_3$/tBuOH	63	98	[38]
3	OH 4-MeOPh–Me	Pd(II)/(-)-sparteine	—	67	99	[34]
		Pd(II)/(-)-sparteine	—	67	98	[37]
		Pd(II)/(-)-sparteine	Cs$_2$CO$_3$/tBuOH	67	99.5	[38]
4	OH	17	(-)-sparteine	62	99	[36]
		Pd(II)/(-)-sparteine	—	69	99.8	[37]
		Pd(II)/(-)-sparteine[c]	Cs$_2$CO$_3$	60	99.6	[39]
		18b	Cs$_2$CO$_3$	64	>99	[45]
		22	—	46[b]	>99	[56]
5	OH tBu–Me	17	(-)-sparteine	61	97	[36]
6	OH 4-BrPh–Me	18b	Cs$_2$CO$_3$	68	93	[45]
7	OH	19b	—	65	>99.5	[47]
8	HO OEt Ph O	VO(OiPr)$_3$/20a	—	51	99	[51]
		Co(OAc)$_2$/25	TEMPO	66	97	[57]
9	HO OiPr iBu O	VO(OiPr)$_3$/20a	—	55	98	[51]
10	HO OMe Ph O	21a	—	54	93	[52]
		21b	—	55	97	[52]
11	Cl OH O	VO(OiPr)$_3$/20b	—	62	99.7	[54]
12	OH	22	—	50[b]	>99	[56]

(continued)

TABLE 4.3 (CONTINUED)
Oxidative Kinetic Resolution of Racemic Secondary Alcohols

N	Substrate	Catalyst	Additive	Conversion (%)	ee (%)	Ref.
13		$Co(OAc)_2$/**25**	TEMPO	61	99.9	[58]

Note: O_2 was used as the oxidant unless otherwise stated.
[a] 1,3-bis(4-bromophenyl)propane-1,3-dione.
[b] Yield of recovered alcohol.

the solvent (the addition of *t*-butanol was no longer required); the process could be performed at 23°C [39]. Palladium(II)–(-)-sparteine-catalyzed oxidative kinetic resolution of alcohols was utilized for the asymmetric syntheses of a series of some key pharmaceuticals such as Prozac, Singulair, the hNK-1 receptor antagonist, and (+)–amurensinine [40,41].

Other chiral ligands were tested in the palladium-catalyzed oxidative kinetic resolutions of secondary alcohols. Stolz and Trend explored the potential of (–)-α-isosparteine and (+)-β-isosparteine: both showed reactivity and enantioselection inferior to (–)-sparteine [42]. A series of (+)-sparteine-like diamines were prepared by O'Brien with co-workers and tested as chiral auxiliaries in the palladium-catalyzed resolution of 1-indanol, demonstrating stereopreference opposite that of (–)-sparteine and inferior enantioselectivities [43].

Sacchetti et al. synthesized a set of novel chiral keto-bispidines that were tested in Pd-catalyzed oxidative kinetic resolution of 1-phenyl ethanol in toluene. The *ees* were in the range of 4 to 42% versus 96% for (–)-sparteine under similar conditions [44]. Shi and co-workers reported the axially chiral palladium complexes of type **18** as efficient catalysts (when used in 10 mol%) for the oxidation of a series of racemic secondary alcohols in toluene [45]. Enantiomeric excesses of 61 to 99% were reported at 61 to 75% conversions [45,46]. The authors screened various basic additives and reported that Cs_2CO_3 was the best and led to the highest alcohol conversions.

Major drawbacks of palladium-catalyzed aerobic oxidative kinetic resolutions of secondary alcohols are low reactivities (requiring reaction times up to 96 hr), and high loads of expensive Pd catalyst (typically 5 to 10 mol%).

Katsuki with co-workers reported chiral ruthenium–salen complexes **19a,b**, capable of oxidative kinetic resolution of secondary alcohols with air oxygen [47]. The reaction proceeded via irradiation with visible light; 2 mol% of the ruthenium catalyst was loaded. Four different alcohols were resolved, and enantiomeric excesses of 62 to >99.5% were reported (Table 4.3). Later, the performance of catalyst **19b** was improved by the addition of 1,3-diketones [48], whereas catalysts of the type **19c** were shown to catalyze the oxidative desymmetrization of 1,4-*meso*-diols [49,50].

Moderate to high *ees* (50-99%) were reported by Toste and co-workers for vanadium-catalyzed oxidative kinetic resolution of α-hydroxy esters with molecular oxygen [51]. The authors used $VO(OiPr)_3$ (5 mol%) as the metal source and

L-t-leucinol-derived Schiff bases of type **20a** as the chirality source, acetone as the solvent, and 3Å molecular sieves. In one case, a very high k_{rel} of >50 was observed [51].

One year later, Chen and co-workers reported a successful use of vanadium complexes **21** with chiral *N*-salicylidene carboxylates for the enantioselective oxidation of α-hydroxy esters and amides under oxygen atmosphere in toluene [52]. They later carried out the reaction in MTBE and considered a broader set of ligand structures and substrates [53].

Li and co-workers tested a series of vanadium(V) complexes generated in situ from VO(O*i*Pr)$_3$ and amino alcohol-derived tridentate Schiff base ligands of type **20** (with R$_1$, R$_2$ = *t*Bu, H, I, various R$_3$, and R$_4$ = H) in the aerobic oxidation of methyl *o*-chloromandelate in acetone [54]. Moderate to high *ee*s for the recovered ester were reported (17 to 99.7%). Jones's group developed a family of polystyrene-supported vanadium catalysts with chiral ligands of type **20** that exhibited good to high enantioselectivities (90 to 99% *ee*) in the oxidative kinetic resolution of ethyl and benzyl mandelates [55].

Other metals are less represented in aerobic oxidative kinetic resolutions of secondary alcohols. Ikariya with co-workers tested a series of chiral half-sandwich iridium, rhodium, and ruthenium complexes of types **22–24** (at 10 mol% loads) for the kinetic resolution of secondary alcohols with air oxygen in THF [56, 57]. Iridium catalysts demonstrated higher k_{rel} values (up to >100 in one case) compared to Rh and Ru compounds.

Sekar and Alamsetti reported the cobalt catalyzed (10 mol% of Co) kinetic resolution of a series of α-hydroxy esters by molecular oxygen, using six different *N*- and/or *O*-functionalized chiral ligands [58]. Chiral ligand **25** was identified as the most efficient, affording enantioenriched α-hydroxy esters with good to high enantioselectivities (78 to 99.9% *ee*). Sekar and Muthupandi reported on the kinetic resolution of racemic α-hydroxy ketones on an in situ-formed chiral Zn-quinine complex in the presence of an oxygen oxidant and TEMPO [59]. They reported low to moderate enantioselectivities (22 to 55% *ee*, and 88% *ee* in one case).

Sekar with co-workers reported the kinetic resolution of racemic benzoins in the presence of in situ–formed chiral iron catalysts. TEMPO was required for achieving acceptable reaction times and enantioselectivities [60]. Katsuki with co-workers studied the iron–salan-catalyzed aerobic kinetic resolution of benzylic secondary alcohols in the presence of 1-naphthol [61]. A binuclear iron complex with salan ligand **26** demonstrated the highest enantioselectivity (up to 99% *ee*).

Overall, transition metal-catalyzed oxidative kinetic resolution of secondary alcohols is less developed than epoxidation and sulfoxidation techniques. We recommend a recent review to interested readers [62].

ENANTIOSELECTIVE AEROBIC OXIDATIVE COUPLING OF 2-NAPHTHOLS

The enantioselective oxidative coupling of 2-naphthols to afford 1,1'-bi-2-naphthols is a challenging route to obtain optically active binols. In this section, several existing catalytic methods of enantioselective oxidative coupling of 2-naphthols with molecular oxygen in the presence of metal complexes are discussed.

Copper complexes play the major role in these transformations. In 1992, Smrčina and Kočovsky with co-workers reported the first catalyzed enantioselective coupling of 2-naphthols in the presence of a copper complex formed in situ from CuCl$_2$ and (–)-sparteine **16** (Figure 4.3) [63]. They synthesized binaphthyl derivatives (S)-(–)-2,2'-dihydroxy-1,1'-binaphthyl, (R)-(+)- and (S)-(–)-2,2'-diamino-1,1'-binaphthyl, and (R)-(+)- and (S)-(–)-2-amino-2'-hydroxy-1,1'-binaphthyl with moderate to good yields, and with up to 100% ee. However, stoichiometric amounts of Cu and sparteine were used, and the enantioselectivity appeared in a combined enantioselective coupling and deracemization process [63].

In a subsequent work, they presented a catalytic version of the Cu–(–)-sparteine system (10 mol%) that allowed the coupling with up to 32% ee using silver(I) chloride as the terminal oxidizing agent (as the reaction should be anaerobic) [64].

The first catalytic enantioselective coupling of 2-naphthol derivatives with O$_2$ as the oxidant was presented by Nakajima and co-workers who adopted L-proline derivatives

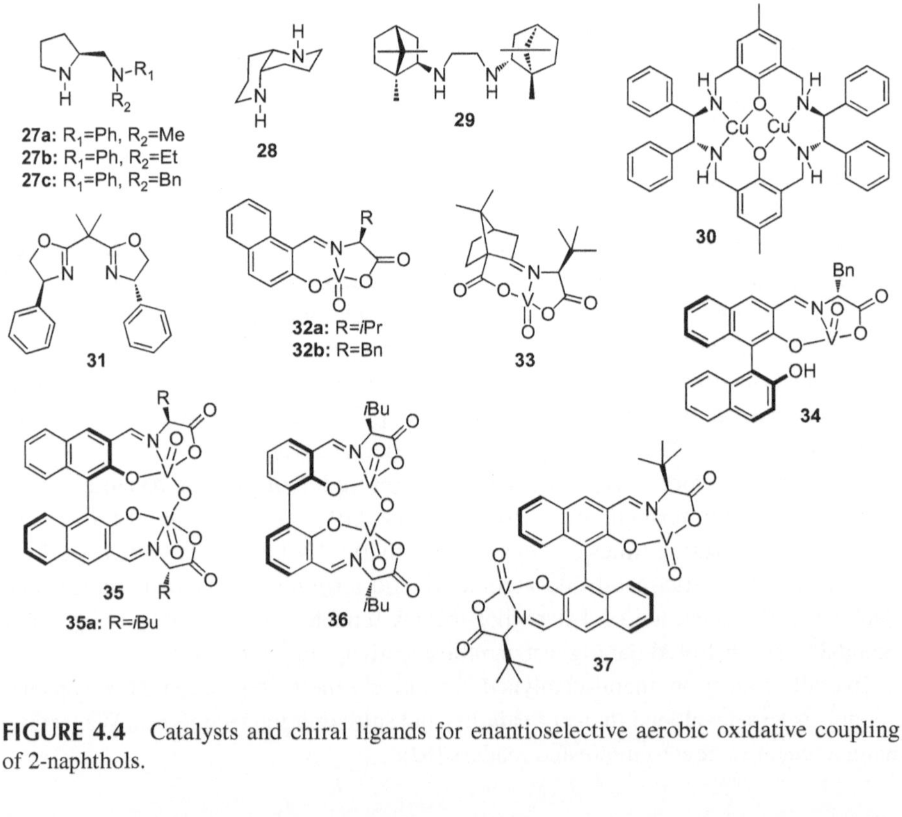

FIGURE 4.4 Catalysts and chiral ligands for enantioselective aerobic oxidative coupling of 2-naphthols.

27 (Figure 4.4) as chiral ligands [65]. Moderate to good yields (24 to 87%) and enantioselectivities (27 to 73% *ee*) were reported for the couplings. The authors found that (–)-sparteine was an inferior asymmetric auxiliary (with respect to diamines **27**). The system was poorly enantioselective for the coupling of naphthols lacking ester moieties at the 3 position [66]. A tenative reaction mechanism was proposed.

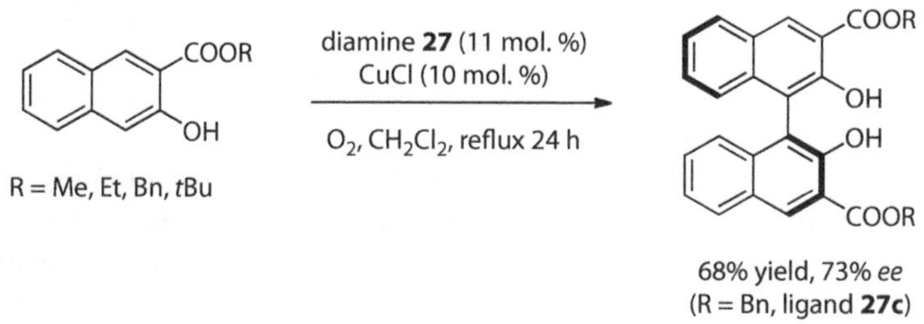

R = Me, Et, Bn, *t*Bu

diamine **27** (11 mol. %)
CuCl (10 mol. %)

O$_2$, CH$_2$Cl$_2$, reflux 24 h

68% yield, 73% *ee*
(R = Bn, ligand **27c**)

Kozlowski and co-workers screened a series of chiral diamines and chose (*S,S*)-1,5-diazadecalin **28** as the most efficient [67]. The authors performed the couplings in toluene without external heating. Reaction times of 3 to 5 days were required. The addition of molecular sieves accelerated the reaction. The method demonstrated high generality for a broad range of substituted 2-naphthols; with good substrates, enantioselectivities up to 96% *ee* were achieved [68].

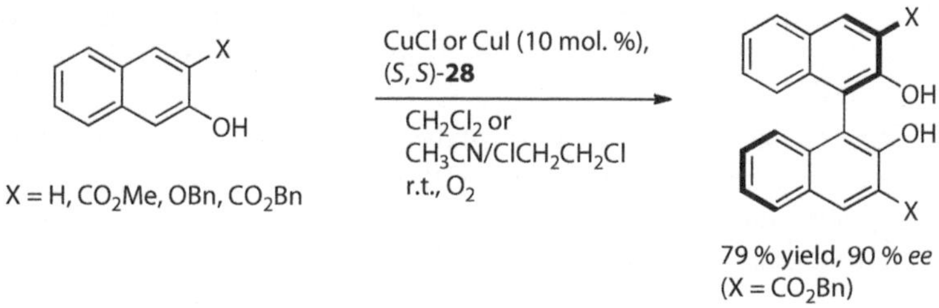

X = H, CO$_2$Me, OBn, CO$_2$Bn

CuCl or CuI (10 mol. %),
(*S, S*)-**28**

CH$_2$Cl$_2$ or
CH$_3$CN/ClCH$_2$CH$_2$Cl
r.t., O$_2$

79 % yield, 90 % *ee*
(X = CO$_2$Bn)

On the basis of mechanistic and stereochemistry studies, a catalytic cycle was proposed [69]. With (+)-camphor-derived diamine **29** as the chiral ligand, Palmisano and Sisti reported lower enantioselectivities (6 to 65% *ee*) in similar Cu-catalyzed couplings [70]. Gao et al. synthesized a series of structurally defined dinuclear copper catalysts; complex **30** catalyzed the homocoupling of 2-naphthol in CCl$_4$ at 0°C with 85% yield and 88% *ee* within 7 days [71].

Little success has been achieved in copper-catalyzed oxidative cross-coupling of two different naphthol derivatives. Habaue with co-workers examined the potential of both enantiomers of *bis*-oxazoline **31** as the chiral ligand in copper-catalyzed oxidative cross-couplings. Various 2-naphthols were cross-coupled in THF under oxygen atmosphere with generally moderate yields and *ees* (10 to 74% *ee*) [72–74]. For the Cu–(–)-sparteine system, poorer enantioselectivities were reported [75].

Ruthenium is another metal that found application in this class of catalyzed transformations. In 2000, Katsuki's group reported the photo-promoted oxidative coupling of 2-naphthol to (*R*)-binol with 72% yield and 65% *ee* in the presence of only 2 mol% of chiral catalyst **19b** (Figure 4.3) [76]. Subsequently, it was shown that 2-naphthols bearing electron-withdrawing substituents (Br and C≡C-Ph) at the 6 positions coupled with slightly higher *ee*s (68-71%) [77].

Several vanadium-based catalyst systems have been reported. The groups of Chen [78,79] and Uang [80,81] studied chiral oxovanadium(IV) complexes of types **32** [78,80], **33** [79], **34** [81]. The complexes were synthesized and studied as catalysts in oxidative homocoupling of substituted naphthols in either CCl$_4$ or CHCl$_3$ at room temperature. The enantioselectivities achieved with **32** and **34** did not exceed 68 and 73% *ee*, respectively; for the **33** catalysts, the coupling of substituted 2-naphthols in up to 87% *ee* was reported.

An attempt to improve the enantioselectivity of vanadium–amino acid-derived Schiff base complexes [78–81] was made by Gong with co-workers. They prepared a series of chiral bimetallic oxovanadium(V) complexes of type **35**, featuring both central chirality stemming from the amino acid and axial chirality from the binol unit [82–84]. Higher enantioselectivities were reported, especially for catalyst **35a** that (at 10 mol% load) catalyzed the coupling of 2-naphthol with to 83% *ee*, and substituted 2-naphthols to 98% *ee*, with good to excellent yields (>99%). Interestingly, catalyst **36**, derived from achiral bisphenol, catalyzed the coupling of 2-naphthol with even higher enantioselectivity (up to 92% *ee*) [83]. Sasai et al. reported a related catalyst **37** without a V-O-V linkage that also appeared to be highly enantioselective (up to 93% *ee*) [85,86].

Katsuki with co-workers reported that a dinuclear iron catalyst with salan ligand **26** (Figure 4.3) catalyzed the highly enantioselective homocoupling [87] and cross-coupling [88] of various 2-naphthols (up to 96% *ee*) and with moderate to good yield (50 to 70%). With 4 mol% of catalyst, the reaction required 48 to 72 hr at 60°C in toluene solution.

The major drawback of metal-based catalyst systems for the oxidative coupling of 2-naphthols is their low reactivity; none of those afforded chiral 1,1′-bi-2-naphthols within acceptable reaction times. Significant improvements are needed before such systems can find practical applications [89,90].

ENANTIOSELECTIVE C–H OXIDATIONS

In recent years, the area of selective (including stereospecific) oxidation of aliphatic C–H groups with H$_2$O$_2$ advanced greatly [91–93]; however, truly enantioselective C–H oxidations (with generation of asymmetric induction in the course of oxidation) remained challenging goals.

In 2000, Ménage with co-workers announced that a dinuclear non-heme iron catalyst **38** (Figure 4.5) demonstrated low enantioselectivities (7 and 15% *ee*, respectively) in the benzylic hydroxylation of ethylbenzene and 1,1′-dimethylindane [94]. However, whether the enantiomeric excess appeared at the C–H oxidation stage or at the expense of subsequent kinetic resolution of the resulting alcohol remains unclear. The second possibility looks plausible because significant amounts of ketone were reported to form (alcohol:ketone ratios of 0.9:1.2 were reported) [94].

In 2012 Simonneaux and co-workers reported a 5% conversion of ethylbenzene to a mixture of (S)-1-phenylethanol and acetophenone (47:53) in the presence of water-soluble iron porphyrin catalyst **39** and H$_2$O$_2$. The resulting 15% *ee* of the formed

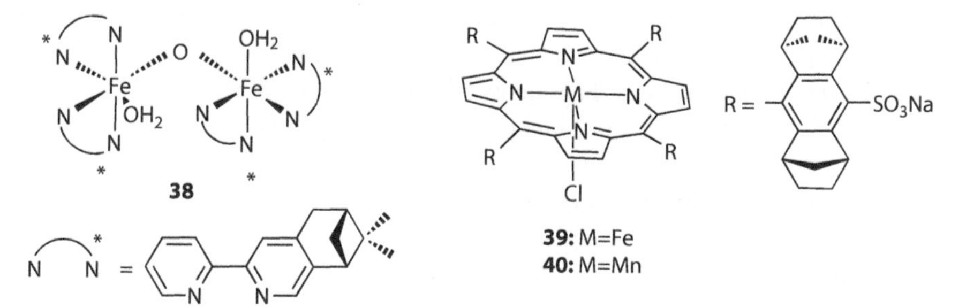

FIGURE 4.5 Catalysts for enantioselective C–H oxidations of alkanes.

alcohol was far more inferior than the result from using iodobenzene diacetate as the oxidant [95].

Simonneaux's group also found that water-soluble manganese–porphyrin complex **40** (2.5 mol%) efficiently catalyzed the oxidation of several alkanes with H_2O_2

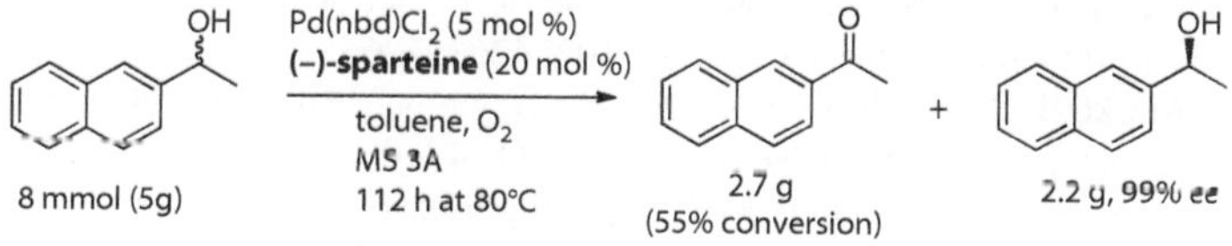

Aerobic oxidative kinetic resolution of 1-naphthalen-2-yl-ethanol [37].

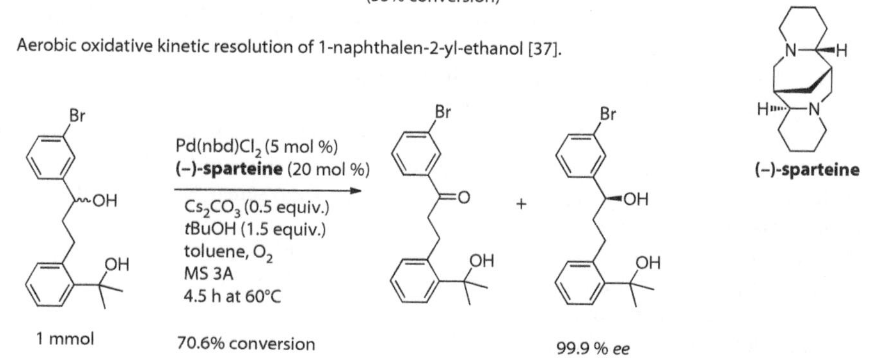

Aerobic oxidative kinetic resolution–key step of the synthesis of leukotriene receptor antagonist Singulair [40].

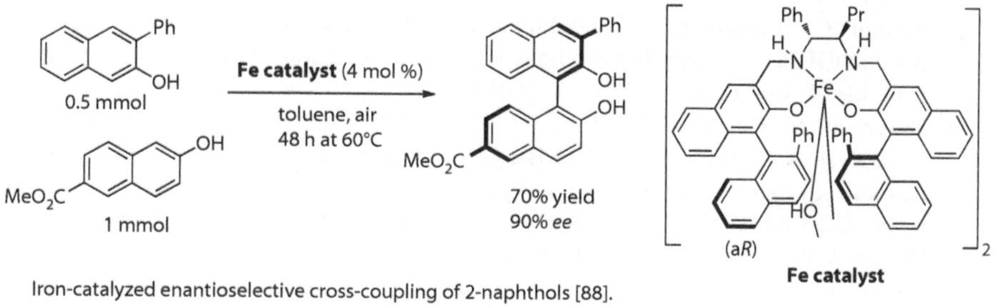

Iron-catalyzed enantioselective cross-coupling of 2-naphthols [88].

FIGURE 4.6 Examples of miscellaneous preparative scale catalytic asymmetric transformations.

(5 equivalents relative to alkane) in water–methanol solutions, with high conversions (88 to 100%) and moderate enantioselectivities (ranging from 32% *ee* for indane to 57% for 4-ethyltoluene) [96]. The alcohol formation selectivity was 93% and this rules out a significant impact of kinetic resolution on hydroxylation enantioselectivity. To date, this has been the only successful example of catalytic enantioselective C–H oxidation with H_2O_2 in the presence of metal complexes.

REFERENCES

1. Bataille, C. J. R. and Donohoe, T. J. 2011. Osmium-free direct syn-dihydroxylation of alkenes. *Chem. Soc. Rev.* 40: 114–128.
2. Johnson, R. A. and Sharpless, K. B. 2000. Catalytic asymmetric dihydroxylation: discovery and development. In *Catalytic Asymmetric Synthesis*, 2nd ed., Ojima, I., Ed. New York: John Wiley & Sons, pp. 357–398.
3. Bolm, C., Hildebrand, J. P., and Muñiz, K. 2000. Recent advances in asymmetric dihydroxylation and aminohydroxylation. In *Catalytic Asymmetric Synthesis*, 2nd ed., Ojima, I., Ed. New York: John Wiley & Sons, Inc., pp. 399–428.
4. Hentges, S. G. and Sharpless, K. B. 1980. Asymmetric induction in the reaction of osmium tetroxide with olefins. *J. Am. Chem. Soc.* 102: 4263–4265.
5. Jacobsen, E. N., Markó, I., Mungall, W. S. et al. 1988. Asymmetric dihydroxylation via ligand-accelerated catalysis. *J. Am. Chem. Soc.* 110: 1968–1970.
6. Costas, M., Tipton, A. K., Chen, K. et al. 2001. Modeling Rieske dioxygenases: the first example of iron-catalyzed asymmetric *cis*-dihydroxylation of olefins. *J. Am. Chem. Soc.* 123: 6722–6723.
7. Suzuki, K., Oldenburg, P. D., and Que, Jr., L. 2008. Iron-catalyzed asymmetric olefin *cis*-dihydroxylation with 97% enantiomeric excess. *Angew. Chem. Int. Ed.* 47: 1887–1889.
8. De Boer, J. W., Browne, W. R., Harutyunyan, S. R. et al. 2008. Manganese-catalyzed asymmetric *cis*-dihydroxylation with H_2O_2. *Chem. Commun.* 3747–3749.
9. Baeyer, A. and Villiger, V. 1899. Einwirkung des Caro'schen Reagens auf Ketone. *Berichte,* 32: 3625-3633.
10. Michelin, R. A., Sgarbossa, P., Scarso, A. et al. 2010. The Baeyer–Villiger oxidation of ketones: paradigm for the role of soft Lewis acidity in homogeneous catalysis. *Coord. Chem. Rev.* 254: 646–660.
11. Mihovilovic, M. D., Rudroff, F., and Grötzl, B. 2004. Enantioselective Baeyer–Villiger oxidations. *Curr. Org. Chem.* 8: 1057–1069.
12. Gusso, A., Baccin, C., Pinna, F. et al. 1994. Platinum-catalyzed oxidations with hydrogen peroxide: enantiospecific Baeyer–Villiger oxidation of cyclic ketones. *Organometallics* 13: 3442–3451.
13. Strukul, G., Varagnolo, A., and Pinna, F. 1997. New (old) hydroxo complexes of platinum(II) as catalysts for the Baeyer–Villiger oxidation of ketones with hydrogen peroxide. *J. Mol. Catal. A Chem.* 117: 413–423.
14. Paneghetti, C., Gavagnin, R., Pinna, F. et al. 1999. New chiral complexes of platinum(II) as catalysts for the enantioselective Baeyer–Villiger oxidation of ketones with hydrogen peroxide: dissymmetrization of *meso*-cyclohexanones. *Organometallics* 18: 5057–5065.
15. Bolm, C., Schlingloff, G., and Weickhardt, K. 1994. Optically active lactones from a Baeyer–Villiger type metal-catalyzed oxidation with molecular oxygen. *Angew. Chem. Int. Ed.* 30: 1848–1849.
16. Bolm, C. and Schlingloff, G. 1995. Metal-catalysed enantiospecific aerobic oxidation of cyclobutanones. *J. Chem. Soc. Chem. Commun.* 1247–1248.

17. Bolm, C., Khan-Luong, T. K., and Schlingloff, G. 1997. Enantioselective metal-catalyzed Baeyer–Villiger oxidation of cyclobutanones. *Synlett.* 1151–1152.

18. Peng, Y., Feng, X., Yu, K. et al. 2001. Synthesis and crystal structure of *bis*(4S,5S)-4,5-dihydro-4,5-diphenyl-2-(2′-oxidophenyl-χO)oxazole-χN.copper(II) and its application in the asymmetric Baeyer–Villiger reaction. *J. Organomet. Chem.* 619: 204–208.

19. Uchida, T. and Katsuki, T. 2001. Cationic Co(III)(salen)-catalyzed enantioselective Baeyer–Villiger oxidation of 3-arylcyclobutanones using hydrogen peroxide as a terminal oxidant. *Tetrahedron Lett.* 42: 6911–6914.

20. Uchida, T. and Katsuki, T. 2002. New asymmetric catalysis by (salen)cobalt(III) complexes (salen = *bis*(salicylidene)ethylenediaminato = {{2,2′-ethane-1,2-diyl.bis(nitrilo-κN)methylidyne.bisphenolato-κO.}(2-)}) of *cis*-β-structure: enantioselective Baeyer–Villiger oxidation of prochiral cyclobutanones. *Helv. Chim. Acta* 85: 3078–3089.

21. Sandaroos, R., Goldani, M. T., Damavandi, S. et al. 2012. Efficient asymmetric Baeyer–Villiger oxidation of prochiral cyclobutanones using new polymer-supported and unsupported chiral Co(salen) complexes. *J. Chem. Sci.* 124: 871–876.

22. Watanabe, A., Uchida, T., Ito, K. et al. 2002. Highly enantioselective Baeyer–Villiger oxidation using Zr(salen) complex as catalyst. *Tetrahedron Lett.* 43: 4481–4485.

23. Matsumoto, K., Watanabe, A., Uchida, T. et al. 2004. Construction of a new asymmetric reaction site: asymmetric 1,4-addition of thiol using pentagonal bipyramidal Hf(salen) complex as catalyst. *Tetrahedron Lett.* 45: 2385–2388.

24. Ito, K., Ishii, A., Kuroda, T. et al. 2003. Asymmetric Baeyer–Villiger oxidation of prochiral cyclobutanones using a chiral cationic Palladium(II) 2-(phosphinophenyl)pyridine complex as catalyst. *Synlett* 643–646.

25. Malkov, A. V., Friscourt, F., Bell, M. et al. 2008. Enantioselective Baeyer–Villiger oxidation catalyzed by palladium(II) complexes with chiral *P,N*-ligands. *J. Org. Chem.* 73: 3996–4003.

26. Cavarzan, A., Bianchini, G., Sgarbossa, P. et al. 2009. Catalytic asymmetric Baeyer–Villiger oxidation in water by using Pt[II] catalysts and hydrogen peroxide: supramolecular control of enantioselectivity *Chem. Eur. J.* 15: 7930–7939.

27. Colladon, M., Scarso, A., and Strukul, G. 2006. Tailoring Pt(II) chiral catalyst design for asymmetric Baeyer–Villiger oxidation of cyclic ketones with hydrogen peroxide. *Synlett.* 3515–3520.

28. Bianchini, G., Cavarzan, A., Scarso, A. et al. 2009. Asymmetric Baeyer–Villiger oxidation with Co(Salen) and H₂O₂ in water: striking supramolecular micelles effect on catalysis. *Green Chem.* 11: 1517–1520.

29. Pazmino, D. E. T., Dudek, H. M., and Fraaije, M. W. 2010. Baeyer–Villiger monooxygenases: recent advances and future challenges. *Curr. Opin. Chem. Biol.* 14: 138–144.

30. Alphand, V. and Wohlgermuth, R. 2010. Applications of Baeyer–Villiger monooxygenases in organic synthesis. *Curr. Org. Chem.* 14: 1928–1965.

31. Terada, M. 2010. Chiral phosphoric acids as versatile catalysts for enantioselective transformations. *Synthesis* 12: 1929–1982.

32. Russo, A., De Fusco, C., and Lattanzi, A. 2012. Organocatalytic asymmetric oxidations with hydrogen peroxide and molecular oxygen. *ChemCatChem* 4: 901–916.

33. Matsumoto, K. and Katsuki, T. 2010. Asymmetric oxidations and related reactions. In *Catalytic Asymmetric Synthesis*, 3rd ed., Ojima, I., Ed. New York: John Wiley & Sons, pp. 839–890.

34. Jensen, D. R., Pugsley, J. S., and Sigman, M. S. 2001. Palladium-catalyzed enantioselective oxidations of alcohols using molecular oxygen. *J. Am. Chem. Soc.* 123: 7475–7476.

35. Mueller, J. A., Jensen, D. R., and Sigman, M. S. 2002. Dual role of (−)-sparteine in the palladium-catalyzed aerobic oxidative kinetic resolution of secondary alcohols. *J. Am. Chem. Soc.* 124: 8202–8203.

36. Mandal, S. K., Jensen, D. R., Pugsley, J. S. et al. 2003. Scope of enantioselective palladium(II)-catalyzed aerobic alcohol oxidations with (–)-sparteine. *J. Org. Chem.* 68: 4600–4603.

37. Ferreira, E. M. and Stolz, B. M. 2001. The palladium-catalyzed oxidative kinetic resolution of secondary alcohols with molecular oxygen. *J. Am. Chem. Soc.* 123: 7725–7726.

38. Bagdanoff, J. T., Ferreira, E. M., and Stolz, B. M. 2003. Palladium-catalyzed enantioselective oxidation of alcohols: a dramatic rate acceleration by Cs_2CO_3/t-BuOH. *Org. Lett.* 5: 835–837.

39. Bagdanoff, J. T. and Stolz, B. M. 2004. Palladium-catalyzed oxidative kinetic resolution with ambient air as the stoichiometric oxidation gas. *Angew. Chem. Int. Ed.* 43: 353–357.

40. Caspi, D. D., Ebner, D. C., Bagdanoff, J. T. et al. 2004. Resolution of important pharmaceutical building blocks by palladium-catalyzed aerobic oxidation of secondary alcohols. *Adv. Synth. Catal.* 346: 185–189.

41. Tambar, U. K., Ebner, D. C., and Stoltz, B. M. 2006. A convergent and enantioselective synthesis of (+)-amurensinine via selective C–H and C–C bond insertion reactions. *J. Am. Chem. Soc.* 128: 11752–11753.

42. Trend, R. M. and Stolz, B. M. 2008. Structural features and reactivity of (sparteine) $PdCl_2$: a model for selectivity in the oxidative kinetic resolution of secondary alcohols. *J. Am. Chem. Soc.* 130: 15957–15966.

43. Dearden, M. J., McGrath, M. J., and O'Brien, P. 2004. Evaluation of (+)-sparteine-like diamines for asymmetric synthesis. *J. Org. Chem.* 69: 5789–5792.

44. Lesma, G., Pilati, T., Sacchetti, A. et al. 2008. New chiral diamino ligands as sparteine analogues: application to the palladium-catalyzed kinetic oxidative resolution of 1-phenyl ethanol. *Tetrahedron Asymmetry* 19: 1363–1366.

45. Chen, T., Jiang, J. J., Xu, Q. et al. 2007. Axially chiral NHC–Pd(II) complexes in the oxidative kinetic resolution of secondary alcohols using molecular oxygen as a terminal oxidant. *Org. Lett.* 9: 865–868.

46. Liu, S. J., Liu, L. J., and Shi, M. 2009. Preparation of novel axially chiral NHC–Pd(II) complexes and their application in oxidative kinetic resolution of secondary alcohols. *Appl. Organomet. Chem.* 23: 183–190.

47. Masutani, K., Uchida, T., Irie, R. et al. 2000. Catalytic asymmetric and chemoselective aerobic oxidation: kinetic resolution of *sec*-alcohols. *Tetrahedron Lett.* 41: 5119–5123.

48. Nakamura, Y., Egami, H., Matsumoto, K. et al. 2007. Aerobic oxidative kinetic resolution of racemic alcohols with bidentate ligand-binding Ru(salen) complex as catalyst. *Tetrahedron* 63: 6383–6387.

49. Shimizu, H., Nakata, K., and Katsuki, T. 2002. (Salen)ruthenium-catalyzed desymmetrization of meso-diols: catalytic aerobic asymmetric oxidation under photoirradiation. *Chem. Lett.* 31: 1080–1081.

50. Shimizu, H. and Katsuki, T. 2003. (Salen)ruthenium-catalyzed desymmetrization of meso-diols (2): apical ligand effect on enantioselectivity. *Chem. Lett.* 32: 480–481.

51. Radosevich, A. T., Musich, C., and Toste, F. D. 2005. Vanadium-catalyzed asymmetric oxidation of r-hydroxy esters using molecular oxygen as stoichiometric oxidant. *J. Am. Chem. Soc.* 127: 1090–1091.

52. Weng, S. S., Shen, M. W., Kao, J. Q. et al. 2006. Chiral N-salicylidene vanadyl carboxylate-catalyzed enantioselective aerobic oxidation of α-hydroxy esters and amides. *Proc. Natl. Acad. Sci. U.S.A.* 103: 3522–3527.

53. Chen, C. T., Kao, J. Q., Salunke, S. B. et al. 2011. Enantioselective aerobic oxidation of α-hydroxy-ketones catalyzed by oxidovanadium(V) methoxides bearing chiral, N-salicylidene-*tert*-butylglycinates. *Org. Lett.* 13: 26–29.

54. Yin, L., Jia, X. A., Li, X. S. et al. 2010. Simply air: vanadium-catalyzed oxidative kinetic resolution of methyl o-chloromandelate by ambient air. *Chin. Chem. Lett.* 21: 774–777.

55. Shiels, R. A., Venkatasubbaiah, K., and Jones, C. W. 2008. Polymer- and silica-supported tridentate Schiff base–vanadium catalysts for asymmetric oxidation of ethyl mandelate: activity, stability and recyclability. *Adv. Synth. Catal.* 350: 2823–2834.

56. Arita, S., Koike, T., Kayaki, Y. et al. 2008. Aerobic oxidative kinetic resolution of racemic secondary alcohols with chiral bifunctional amido complexes. *Angew. Chem. Int. Ed.* 47: 2447–2449.

57. Ikariya, T., Kuwata, S., and Kayaki, Y. 2010. Aerobic oxidation with bifunctional molecular catalysts. *Pure Appl. Chem.* 82: 1471–1483.

58. Alamsetti, S. K. and Sekar, G. 2010. Chiral cobalt-catalyzed enantioselective aerobic oxidation of α-hydroxy esters. *Chem. Commun.* 7235–7237.

59. Muthapandi, P. and Sekar, G. 2011. Chiral Zn-catalyzed aerobic oxidative kinetic resolution of α-hydroxy ketones. *Tetrahedron Asymmetry.* 22: 512–517.

60. Muthapandi, P., Kumar, S., and Sekar, G. 2009. Chiral iron complex catalyzed enantioselective oxidation of racemic benzoins. *Chem. Commun.* 3288–3290.

61. Kunisu, T., Oguma, T., and Katsuki, T. 2011. Aerobic oxidative kinetic resolution of secondary alcohols with naphthoxide-bound iron(salan) complex. *J. Am. Chem. Soc.* 133: 12937–12939.

62. Parmeggiani, C. and Cardona, F. 2012. Transition metal-based catalysts in the aerobic oxidation of alcohols. *Green Chem.* 14: 547–564.

63. Smrčina, M., Lorenc, M., Hanuš-Sedmera, P., and Kočovsky, P. 1992. Synthesis of enantiomerically pure 2,2'-dihydroxy- 1,1'-binaphthyl, 2,2'-diamino-1,1'-binaphthyl, and 2-amino-2'-hydroxy-1,1'-binaphthyl: comparison of processes operating as diastereoselective crystallization and as second-order asymmetric transformation. *J. Org. Chem.* 57: 1917–1920.

64. Smrčina, M., Polákova, J., Vyskočil, Š. et al. 1993. Synthesis of enantiomerically pure binaphthyl derivatives: mechanism of the enantioselective oxidative coupling of naphthols and designing a catalytic cycle. *J. Org. Chem.* 58: 4534–4538.

65. Nakajima, M., Kanayama, K., Miyoshi, I. et al. 1995. Catalytic asymmetric synthesis of binaphthol derivatives by aerobic oxidative coupling of 3-hydroxy-2-naphthoates with chiral diamine–copper complex. *Tetrahedron Lett.* 36: 9519–9520.

66. Nakajima, M., Miyoshi, I., Kanayama, K. et al. 1999. Enantioselective synthesis of binaphthol derivatives by oxidative coupling of naphthol derivatives catalyzed by chiral diamine–copper complexes. *J. Org. Chem.* 64: 2264–2271.

67. Li, X., Yang, J., and Kozlowski, M. C., 2001. Enantioselective oxidative biaryl coupling reactions catalyzed by 1,5-diazadecalin metal complexes. *Org. Lett.* 3: 1137–1140.

68. Li, X., Hewgley, B., Mulrooney, C. A. et al. 2003. Enantioselective oxidative biaryl coupling reactions catalyzed by 1,5-diazadecalin metal complexes: efficient formation of chiral functionalized binol derivatives. *J. Org. Chem.* 68: 5500–5511.

69. Hewgley, J. B., Stahl, S. S., and Kozlowski, M. C. 2008. Mechanistic study of asymmetric oxidative biaryl coupling: evidence for self-processing of copper catalyst to achieve control of oxidase versus oxygenase activity. *J. Am. Chem. Soc.* 130: 12232–12233.

70. Caselli, A., Giovenzana, G. B., Palmisano, G. et al. 2003. Synthesis of C2–symmetrical diamine based on (1R)-(+)-camphor and application to oxidative aryl coupling of naphthols. *Tetrahedron Asymmetry* 14: 1451–1454.

71. Gao, J., Reibenspies, J. H., and Martell, A. E. 2003. Structurally defined catalysts for enantioselective oxidative coupling reactions. *Angew. Chem. Int. Ed.* 42: 6008–6012.

72. Temma, T. and Habaue, S. 2005. Highly selective oxidative cross-coupling of 2-naphthol derivatives with chiral copper(I)–bisoxazoline catalysts. *Tetrahedron Lett.* 46: 5655–5657.

73. Temma, T., Hatano, B., and Habaue, S. 2006. Cu(I)-catalyzed asymmetric oxidative cross-coupling of 2-naphthol derivatives. *Tetrahedron* 62: 8559–8563.

74. Temma, T., Hatano, B., and Habaue, S. 2006. Copper(I) catalyzed asymmetric oxidative cross-coupling copolymerization leading to alternating copolymers. *Polymer* 47: 1845–1851.
75. Yusa, Y., Kaito, I., Akiyama, K. et al. 2010. Asymmetric catalysis of homo-coupling of 3-substituted naphthylamine and hetero-coupling with 3-substituted naphthol leading to 3,3′-dimethyl-2,2′-diaminobinaphthyl and -2-amino-2′-hydroxybinaphthyl. *Chirality* 2: 224–228.
76. Irie, R., Masutani, K., and Katsuki, T. 2000. Asymmetric aerobic oxidative coupling of 2-naphthol derivatives catalyzed by photo-activated chiral (NO)Ru(II)–salen complex. *Synlett.* 1433–1436.
77. Irie, R. and Katsuki, T. 2004. Selective aerobic oxidation of hydroxy compounds catalyzed by photoactivated ruthenium–salen complexes (selective catalytic aerobic oxidation). *Chem. Rec.* 4: 96–109.
78. Hon, S. W., Li, C. H., Kuo, J. H. et al. Catalytic asymmetric coupling of 2-naphthols by chiral tridentate oxovanadium(IV) complexes. *Org. Lett.* 3: 869–872.
79. Barhate, N. B. and Chen, C. T. 2002. Catalytic asymmetric oxidative couplings of 2-naphthols by tridentate N-ketopinidene-based vanadyl dicarboxylates. *Org. Lett.* 4: 2529–2532.
80. Chu, C. Y., Hwang, D. R., Wang, S. K. et al. 2001. Chiral oxovanadium complex catalyzed enantioselective oxidative coupling of 2-naphthols. *Chem. Commun.* 980–981.
81. Chu, C. Y. and Uang, B. J. 2003. Catalytic enantioselective coupling of 2-naphthols by new chiral oxovanadium complexes bearing a self accelerating functional group. *Tetrahedron Asymmetry* 14: 53–55.
82. Luo, Z., Gong, L., Cui, X. et al. 2002. The rational design of novel chiral oxovanadium(IV) complexes for highly enantioselective oxidative coupling of 2-naphthols. *Chem. Commun.* 914–915.
83. Luo, Z., Liu, Q., Gong, L. et al. 2002. Novel achiral biphenol-derived diastereomeric oxovanadium(iv) complexes for highly enantioselective oxidative coupling of 2-naphthols. *Angew Chem. Int. Ed.* 41: 4532–4535.
84. Guo, Q. X., Wu, Z. J., Luo, Z. B. et al. 2007. Highly enantioselective oxidative couplings of 2-naphthols catalyzed by chiral bimetallic oxovanadium complexes with oxygen or air as oxidant. *J. Am. Chem. Soc.* 129: 13927–13938.
85. Somei, H., Asano, Y., Yoshida, T. et al. 2004. Dual activation in a homolytic coupling reaction promoted by an enantioselective dinuclear vanadium(IV) catalyst. *Tetrahedron Lett.* 45: 1841–1844.
86. Takizawa, S., Rajesh, D., Katayama, T. et al. 2009. One-pot preparation of chiral dinuclear vanadium(V) complex. *Synlett* 1667–1669.
87. Egami, H. and Katsuki, T. 2009. Iron-catalyzed asymmetric aerobic oxidation: oxidative coupling of 2-naphthols. *J. Am. Chem. Soc.* 132: 6082–6083.
88. Egami, H., Matsumoto, K., Oguma, T. et al. 2010. Enantioenriched synthesis of C_1-symmetric binols: iron-catalyzed cross-coupling of 2-naphthols and some mechanistic insight. *J. Am. Chem. Soc.* 132: 13633–13635.
89. Giri, R., Shi, B. F., Engle, K. M. et al. 2009. Transition metal-catalyzed C–H activation reactions: diastereoselectivity and enantioselectivity. *Chem. Soc. Rev.* 38: 3242–3272.
90. Wang, H. 2010. Recent advances in asymmetric oxidative coupling of 2-naphthol and its derivatives. *Chirality* 22: 827–837.
91. Talsi, E. P. and Bryliakov, K. P. 2012. Chemo- and stereoselective C–H oxidations and epoxidations/cis-dihydroxylations with H_2O_2, catalyzed by non-heme iron and manganese complexes. *Coord. Chem. Rev.* 256: 1418–1434.
92. Company, A., Lloret, J., Gómez, L. et al. 2012. Alkane C–H oxygenation catalyzed by transition metal complexes. In *Catalysis by Metal Complexes*, Vol. 38, P. J. Pérez, Ed. Dordrecht: Springer, pp. 143–228.

93. Roduner, E., Kaim, W., Sarkar, B. et al. 2013. Selective catalytic oxidation of C–H bonds with molecular oxygen. *ChemCatChem.* 5: 82–112.
94. Mekmouch, Y., Duboc-Toia, C., Ménage, S. et al. 2000. Hydroxylation of alkanes catalysed by a chiral m-oxo diferric complex: a metal-based mechanism. *J. Mol. Catal. A Chem.* 156: 85–89.
95. Le Maux, P., Srour, H. F., and Simonneaux, G. 2012. Enantioselective water-soluble iron–porphyrin-catalyzed epoxidation with aqueous hydrogen peroxide and hydroxylation with iodobenzene diacetate. *Tetrahedron* 68: 5824–5828.
96. Srour, H. F., Le Maux, P., and Simonneaux, G. 2012. Enantioselective manganese–porphyrin-catalyzed epoxidation and C–H hydroxylation with hydrogen peroxide in water and methanol solutions. *Inorg. Chem.* 51: 5850–5856.

5 Organocatalytic Asymmetric Oxidations

In recent decades, asymmetric organocatalysis (including organocatalytic asymmetric oxidations) emerged as a challenging area of modern catalytic chemistry and provided powerful synthetic tools for metal-free catalytic procedures [1–7]. The attention paid to organocatalytic synthetic methods continues to increase because of stricter safety, selectivity, and sustainability restrictions.

The advantages of organocatalysts over conventional metallocomplex catalysts are the availability and low cost of raw materials (from the chiral pool or its simple derivatives), stability and non-toxicity, and good moisture and air tolerance characteristics [1–3,8]. Since transition metals are not involved (organocatalysts are composed of carbon, hydrogen, nitrogen, sulfur, and phosphorus), organocatalytic methods are especially attractive when preparing compounds such as pharmaceuticals for that hard constraints on residual metal contamination are the case [1].

In the area of organocatalytic asymmetric oxidations, various techniques for the epoxidation of unfunctionalized olefins and electron-deficient α,β-unsaturated ketones and aldehydes, oxidation of sulfides, and Baeyer–Villiger oxidations with H_2O_2 (or its derivatives) and with O_2 have been reported. This chapter reviews these catalytic processes.

EPOXIDATIONS

In recent years, several specialized and comprehensive reviews on enantioselective organocatalytic epoxidations have appeared [4–11]. Most organocatalytic methods focus on the epoxidation of α,β-unsaturated ketones (α,β-enones) [4]. The three major classes of organocatalytic processes used in asymmetric epoxidations are: (1) those exploiting either chiral ketones or iminium salts as catalysts; (2) those relying on polypeptide catalysts; and (3) on-phase transfer catalysts.

CHIRAL KETONE– AND IMINIUM SALT–CATALYZED EPOXIDATIONS

In the course of ketone or iminium salt–catalyzed oxidations, the intermediate chiral dioxiranes or oxaziridinium cations are formed that further transfer an oxygen atom to the olefinic double bond in a stereoselective fashion [2]. Until 1999, only oxone (potassium persulfate $KHSO_5$ in the form of its triple salt with K_2SO_4 and $KHSO_4$) acted as a terminal oxidant in such reactions.

The use of hydrogen peroxide for this purpose was first reported in 1999 when Shu and Shi demonstrated that in the presence of K_2CO_3, H_2O_2 can react with the reaction solvent (acetonitrile) to form peroxyimidic acid that converts the ketone to the

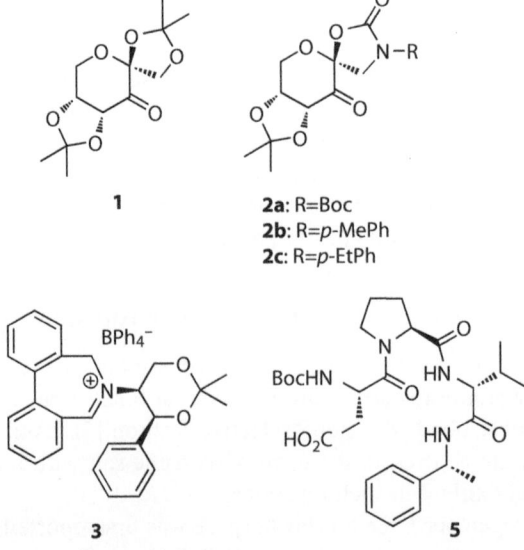

desired dioxirane [12,13]. The authors used a fructose-derived ketone **1** (Figure 5.1) as the catalyst (with 10 to 30 mol% load) and 3 to 4 equivalents of H_2O_2, and aqueous K_2CO_3 as additives [12,13]. For five di- and trisubstituted *E*-olefins, enantioselectivities of 89 to 95% *ee* were reported (Table 5.1) [12]. In a subsequent work, the authors considered a broader range of substrates, and reported enantioselectivities up to 99% *ee* in one case [13].

Later, Shi and co-workers found that oxazolidinone-containing ketones of type **2** could also catalyze the epoxidation of *cis*- and terminal olefins by H_2O_2 with high enantioselectivities in the presence of CH_3CN [14]. Crucially, *n*-butanol was identified as the optimal solvent; 3.0 equivalents of H_2O_2 and 3.8 of CH_3CN were employed as the terminal oxidant. Chiral catalyst **1** was applied for the enantioselective

FIGURE 5.1 Ketone and iminium salt-based catalysts for enantioselective epoxidation of olefins with H_2O_2.

TABLE 5.1
Asymmetric Epoxidation of Olefins with H_2O_2 Catalyzed by Chiral Ketones

N	Substrate	Catalyst	Oxidant	Additive	Epoxide (%)	ee (%)	Ref.
1		1	H_2O_2	K_2CO_3	84	92	[12]
		1	H_2O_2	K_2CO_3	94	93	[13]
2		1	H_2O_2	K_2CO_3	94	95	[13]
		2c	H_2O_2	K_2CO_3	78	88	[14]
3		1	H_2O_2	K_2CO_3	78	99	[13]
4		1	H_2O_2	K_2CO_3	94	98	[13]
		3	H_2O_2	$NaHCO_3$	a	46	[16]
5		2c	H_2O_2	K_2CO_3	83	82	[14]
6		2c	H_2O_2	K_2CO_3	93	83	[14]
7		2c	H_2O_2	K_2CO_3	89	91	[14]

a Quantitative conversion of 1-phenylcyclohexene.

epoxidation of *cis*-1-propenylphosphonic acid, a precursor of the corresponding epoxide, broad-spectrum antibiotic fosfomycin, (with 100% conversion and up to 68% *ee*) [15].

Until recently, there have been no examples of iminium-catalyzed alkene epoxidations by hydrogen peroxide; apparently, the latter could not directly convert the iminium cation into the corresponding oxaziridinium one. Buckley with co-workers developed a procedure based on the use of proper co-catalysts (carbonates and hydrocarbonates): the latter aided in transferring oxygen from 50% H_2O_2 (6 equivalents) to the iminium salt catalyst **3** (5 to 10 mol%). Moderate *ees* were obtained in the epoxidation of 1-phenylcyclohexene (15 to 46% *ee*) in a CH_3CN–water medium [16].

EPOXIDATION OF α,β-UNSATURATED KETONES

In contrast to ketone and iminium salt-catalyzed enantioselective epoxidation of olefins with H_2O_2, polypeptide-catalyzed epoxidations of electron-deficient olefins (such as α,β-unsaturated ketones) with hydrogen peroxide is a mature technique developed in the early 1980s when Juliá and Colonna reported the epoxidation of chalcone (and other electron-poor olefins) with basic hydrogen peroxide in the presence of poly-*L*-alanine **4a** at 85% yield and 93% *ee* [17–21].

4a: R=H
4b: R=*i*Pr

The catalytic reaction was simple (room temperature, ambient atmosphere) and highly enantioselective. After optimization, the *ee*s improved to 96% with poly-*L*-alanine **4a** and to 97% with poly-*L*-leucine **4b** [21]. Major disadvantages of the reported "synthetic enzyme" systems were the high excess of oxidant (generally more than 10 equivalents), the need for pre-activation time, poor catalyst recyclability, unacceptably long reaction times (≥ 3 days), and gradual degradation of the solid catalyst under basic conditions.

Nevertheless, the Juliá-Colonna polypeptide-catalyzed method for the epoxidation of enones demonstrated great practical potential and inspired a number of related works by other groups that broadened the substrate scope and introduced several useful improvements. In addition to a variety of enones, the method demonstrated its applicability for the epoxidation of enynones, enediones, unsaturated ketoesters [22], and geminally disubstituted and trisubstituted enones [23].

In a number of papers, the original three-phase (toluene (or CH_2Cl_2)–aqueous H_2O_2–solid catalyst) reaction system was replaced with a two-phase system: polar organic solvent (THF, *t*BuOMe, EtOAc, DME) and solid catalyst, using 1.2 to 4.8 equivalents of a solid H_2O_2-containing oxidant (urea hydroperoxide or $Na_2CO_3 \cdot 1.5H_2O_2$) as the oxygen source [24–29]. Two-phase systems demonstrated higher reactivities, allowing dramatically shorter reaction times.

Degradation of the catalyst was reduced to a minimum by replacing aqueous NaOH with a non-nucleophilic base 1,8-diazabicyclo[5.4.0]undec-7-ene (DBU). To improve the recyclability, the polypeptide catalysts were supported on solid supports [25–27,29–33]. Remarkably, the use of sodium percarbonate as the oxygen source did not require the addition of expensive DBU as an external base [27 28,31,32].

Geller with co-workers developed a novel high-temperature polypeptide preparation procedure. The prepared catalysts did not need prolonged pre-activation periods and demonstrated very high activity levels (tetrabutylammonium bromide served as phase transfer agent). The reaction required only 7 min under triphasic conditions (in the presence of 28.5 equivalents of basic H_2O_2), with retention of high enantioselectivity and quantitative conversion [34,35]. The authors also found that the amount of oxidant (H_2O_2) could be reduced from 28.5 equivalents to only 1.3 [36] and developed a scaled-up procedure applicable to hundred-gram substrate loads [37].

More recently, Tang with co-workers reported the preparation of an imidazolium-modified poly-*L*-leucine compound that catalyzed the enantioselective epoxidations (with 1.5 equivalents of aqueous sodium percarbonate, $Na_2CO_3 \cdot 1.5H_2O_2$) without pre-activation and was recovered easily by filtration and reused at least seven times without loss in enantioselectivity [38].

The groups of Roberts and Berkessel studied the effect of the primary structure of the polypeptide catalyst and its helicity on epoxidation enantioselectivity [39–42]

TABLE 5.2
Polypeptide-Catalyzed Asymmetric Epoxidation of α,β-Unsaturated Ketones with H_2O_2 and Its Derivatives

N	Substrate	Catalyst	Oxidant	Additive	Epoxide (%)	ee (%)	Ref.
1		poly-L-alanine	H_2O_2	NaOH	85	93	[17]
		poly-L-leucine	H_2O_2	NaOH	80	97	[21]
		poly-L-leucine	UHP	DBU	85	>95	[24]
		poly-L-leucine/CLAMPS	UHP	DBU	93	96	[26]
		poly-L-leucine/CLAMPS	$Na_2CO_3 \cdot 1.5H_2O_2$	—	>99	96	[27]
		poly-L-leucine/CLAMPS	$Na_2CO_3 \cdot 1.5H_2O_2$	—	95[a]	95	[31]
		poly-L-leucine/silica	$Na_2CO_3 \cdot 1.5H_2O_2$	—	94	93	[32]
		oligo-L-leucine/PEG	UHP	DBU	80[a]	98	[44]
		oligo-L-leucine/Ps	UHP	NaOH	92[a]	97	[45]
2		poly-L-leucine	H_2O_2	NaOH	92	>98	[22]
		poly-L-leucine[b]	UHP	DBU	99[a]	97	[28]
3		poly-L-leucine	H_2O_2	NaOH	>95	>95	[22]
4		poly-L-leucine	H_2O_2	NaOH	60	62	[24]
		poly-L-leucine/silica	UHP	DBU	78	93	[25]
		PLN[c]/silica	UHP	DBU	88	92	[31]

(continued)

TABLE 5.2 (CONTINUED)

Polypeptide-Catalyzed Asymmetric Epoxidation of α,β-Unsaturated Ketones with H_2O_2 and Its Derivatives

N	Substrate	Catalyst	Oxidant	Additive	Epoxide (%)	ee (%)	Ref.
5		poly-L-leucine/PSDVB[d]	H_2O_2	NaOH	98	99	[30]
6		poly-L-leucine	H_2O_2	NaOH[e]	78[f]	97.3	[37]
7		poly-L-leucine[g]	$Na_2CO_3 \cdot 1.5H_2O_2$	—	98	99	[38]
8	**5**	**5**	H_2O_2	DIC, DMAP[h]	76	92	[49]

a   Conversion reported.
b   Diaminopropane-bound poly-L-leucine.
c   Poly-L-neo-pentylglycine.
d   Poly(styrene-co-divinylbenzene)-supported poly-L-leucine.
e   With phase-transfer catalysis tetrabutylammonium bromide.
f   Reaction performed at 100 g substrate load.
g   Imidazolium-modified.
h   Diisopropylcarbodiimide (DIC), 2.0 equivalents; DMAP = N,N-dimethyl-4-aminopyridine (phase-transfer agent).

and concluded that the enone substrate preferentially binds to the polyleucine near the N-terminus [40–42]. This indicates that even very short polypeptides should afford epoxides with excellent enantioselectivities and thus opens the gate to replacing solid polypeptide bulk with shorter oligopeptides. Interesting embodiments for these considerations were reported; in particular, Okhata with co-workers prepared a series of oligoleucines (soluble in a variety of organic solvents) and showed that they catalyzed the epoxidation of chalcone with UHP and DBU up to 94% *ee* [43].

Another approach relied on polyethylene glycol- [44–46] and polystyrene-bound oligo-L-leucines [45] as catalysts. The resulting homogeneous catalyst systems, soluble in THF, in some cases demonstrated higher catalytic activities (with reaction times as short as 15 to 60 min) [45] compared to several days for the original heterogeneous Juliá-Colonna epoxidations), using UHP as the oxidant and DBU [40] or NaOH [45] as the base. In the latter case, a continuous flow reaction setup was developed. The catalyst was retained in the reactor by a nanofiltration membrane and was used up to 25 retention times without significant loss of enantioselectivity (see also Chapter 8).

To date, Juliá-Colonna epoxidations are regarded as one the most efficient organocatalytic routes to α,β-epoxyketones [5,6]. Various modifications of Juliá-Colonna procedures have been applied to synthesize precursors of various biologically active organic compounds, such as (+)-clausenamide [26], as diltiazem and Taxol side chain [47], naturally occurring styryl lactones [48], and others [6].

An interesting complement to the polypeptide catalysts is a tripeptide system that may form an in situ peracid moiety. In particular, peptide **5** (5 to 10 mol%) was shown to catalyze the epoxidation of 1-substituted cylcohexenones and other alkenes with H_2O_2 (2.5 equivalents) in a CH_2Cl_2 and H_2O medium up to 92% *ee* [49]. To facilitate the process, a phase transfer catalyst (DMAP) was added; however, in most cases the reaction took 1 to 3 days. In a subsequent publication, the authors disclosed the crucial functional role of the Pro-D-Val moiety. Replacement of the latter with an isosteric alkene dramatically reduced the epoxidation enantioselectivity [50]. More recently, catalysts of type **5** demonstrated high selectivities in the enantio- and diastereoselective oxidation of various indole derivatives [51].

Another well-developed approach, entirely focused on the enantioselective epoxidation of α,β-unsaturated ketones, exploits phase transfer organic catalysis. In 1976, Wynberg and co-workers pioneered the asymmetric epoxidation of several enones (*trans*-chalcones and naphthoquinones) with 30% H_2O_2 in toluene with moderate enantioselectivities (up to 25% *ee*), using 2 mol% of cinchona alkaloid-derived quaternary ammonium chloride salt **6** (Figure 5.2) [52]. They used a biphasic system (aqueous H_2O_2 and CCl_4) and added NaOH as a basic co-catalyst. In subsequent communications, the authors reported improved enantioselectivities up to 55% *ee* [53–57].

Much later, Arai and Shiori and co-workers screened a number of cinchona catalysts of type **7** (5 mol%) bearing electron-withdrawing substituents at the *p*-position of the aryl substituent R, and suggested adding LiOH that yielded better *ee*s than NaOH [58,59]. Substituted chalcones were epoxidized in up to 92% *ee* (Table 5.3). Interestingly, while the oxidation of *trans*-enones in most cases proceeded more enantioselectively in *n*-Bu$_2$O, CHCl$_3$ was the organic solvent of choice for the epoxidation of substituted naphthoquinones [58,59].

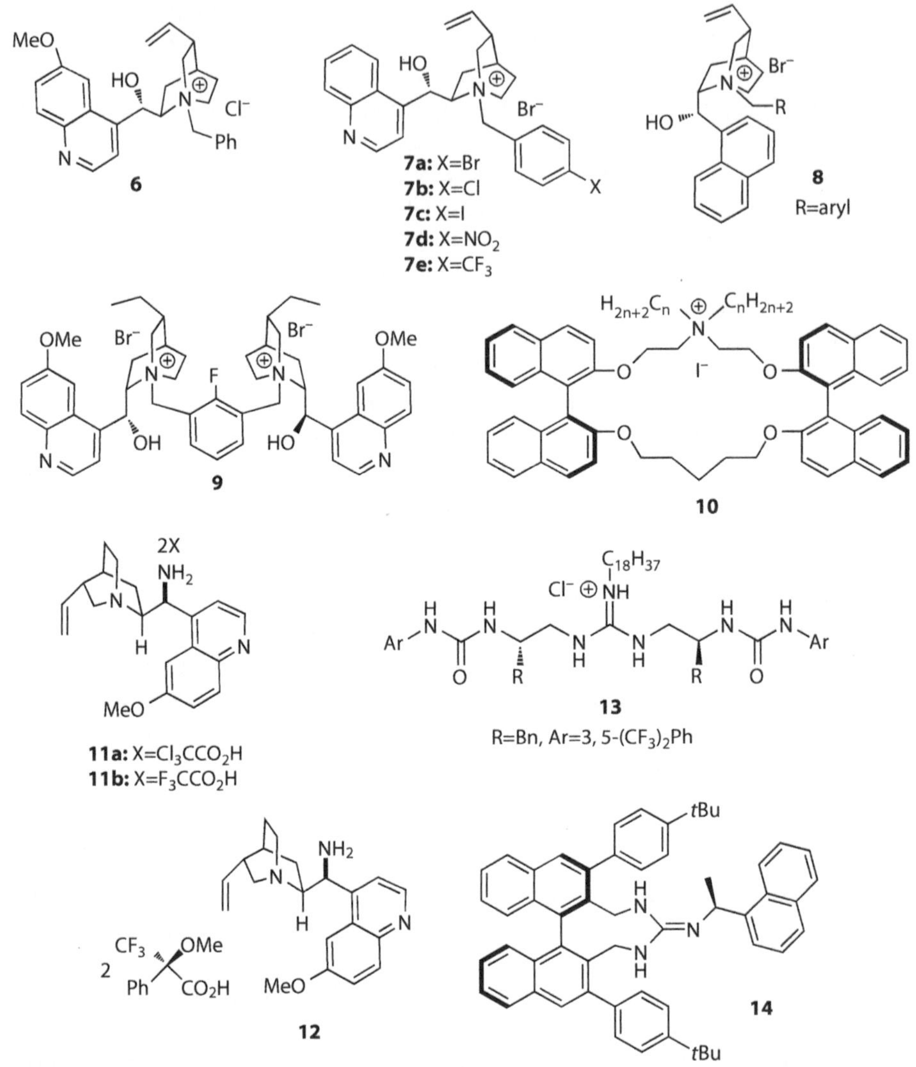

FIGURE 5.2 Phase transfer organic catalysts for enantioselective epoxidation of α,β-unsaturated ketones with H$_2$O$_2$.

Dehmlow and co-workers demonstrated that quaternary salts of structural analogues of cinchona alkaloids of type **8** could also efficiently catalyze the enantioselective epoxidation of 2-isopropyl-1,4-naphthoquinone with moderate to good enantioselectivities (up to 84% *ee*) [60]. Moderate enantioselectivities (up to 61% *ee*) were reported by Gao with co-workers for the epoxidation of enones on phase transfer catalysts prepared from cinchona alkaloids and Fréchet dendritic wedges [61].

An important improvement introduced by Jew et al. consisted in the addition of a commercially available surfactant, like Triton X-100 or Span 20 (sorbitan monolaurate), into the reaction mixture. In effect, the dimeric cinchona phase transfer

TABLE 5.3

Asymmetric Epoxidation of α,β-Unsaturated Ketones with H_2O_2 Mediated by Cinchona Alkaloid-Derived and Related Phase Transfer Catalysts

N	Substrate	Catalyst	Oxidant	Additive	Epoxide (%)	ee (%)	Ref.
1		6	H_2O_2	NaOH	92	48	[54]
2		6	H_2O_2	NaOH	92	34	[52]
		7c	H_2O_2	LiOH	97	84	[58]
		9	H_2O_2	KOH, Span 20	95	>99	[62]
		13	H_2O_2	NaOH	99	94	[67]
3		7c	H_2O_2	LiOH	100	92	[59]
		9	H_2O_2	KOH, Span 20	95	97	[62]
4		8	H_2O_2	LiOH	75	84	[60]
5	2-naphthyl	10 (n = 9)	H_2O_2	KOH	90	83	[63]
		13	H_2O_2	NaOH	98	96	[67]
6	C_9H_{19}	11b	H_2O_2	—	82	98	[64]
7	C_5H_{11}	11b	H_2O_2	—	81	97	[64]
8		11b	H_2O_2	—	58	94	[65]
9		12	H_2O_2	—	87	90	[66]

catalyst **9**, used in only 1 mol%, afforded epoxides of chalcone derivatives in very high yields and *ee*s [62].

Besides the cinchona alkaloid-derived phase transfer catalysts, binol-derived quaternary ammonium salts **10** were reported to catalyze the epoxidation of chalcones with basic hydrogen peroxide, showing moderate *ee*s; the enantioselectivity level could be controlled by the lengths of the alkyl chains at the quarternary nitrogen atom [63].

List and co-workers reported the use of cinchona alkaloid-derived amine catalysts of type **11**. In the form of trichloro- or trifluoroacetic acid salt, they showed high enantioselectivities (over 90% *ee* in two cases) in the epoxidation of simple acyclic aliphatic α,β-unsaturated ketones [64]. The authors suggested a plausible catalytic cycle, including the basic hydrolysis of an intermediate peroxyhemiketal.

The same group extended their studies on a series of salts of chiral amines and diamines with chiral acids and reported the highly enantioselective epoxidation of several cyclic enones (mainly substituted cyclohexenones) with 1.5 equivalents of H$_2$O$_2$ [65]. Later, the same group reported the replacement of trichloro- or trifluoro-acetic acid with (R)-Mosher's acid that led to catalyst 12 that was highly enantiose-lective for the epoxidation of cyclopentenones [66].

An interesting system was presented by Tanaka and Nagasawa who prepared a guanidine–urea bifunctional organocatalyst of type 13 and conducted the successful epoxidation of chalcone derivatives with basic H$_2$O$_2$ with high enantioselectivities (up to 96% ee) [67]. Terada and Nakano used sterically encumbered binol derivatives of type 14 as catalysts and conducted the epoxidation of several α,β-enones with moderate enantioselectivities [68].

To date, the epoxidation of α,β-unsaturated ketones under phase transfer catalytic conditions has been more stereoselective with sodium hypochlorite as the oxidant; hydrogen peroxide demonstrated inferior results. Other significant drawbacks are the high excess of H$_2$O$_2$ required (typically 5 to 30 equivalents) and long reaction times (hours to days) that have limited practical applications of such catalyst systems to date.

EPOXIDATION OF α,β-UNSATURATED ALDEHYDES

The α,β-unsaturated aldehydes represent another challenging class of substrates for asymmetric epoxidation. In 2005, the first organocatalytic experimental procedure for the enantioselective epoxidation of α,β-unsaturated aldehydes was reported [69]. In the presence of 10 mol% of chiral amine catalyst 15a (Figure 5.3) and 1.3 equiva-lents of H$_2$O$_2$, the reaction proceeded to full conversion within a few hours at room temperature in CH$_2$Cl$_2$, affording epoxides with up to 98% ees (Table 5.4).

UHP could also be used as an oxidant to generate similarly high ees. With greater excess of H$_2$O$_2$ (3 equivalents), the system also performed satisfactorily in EtOH:H$_2$O (3:1), demonstrating somewhat lower epoxide yields and enantioselec-tivities [70]. Córdova and co-workers screened a library of chiral proline derivatives and identified compounds 15b and 16 as ensuring the highest enantioselectivities (97 to 98% ee) in the epoxidation of cinnamic and other α,β-unsaturated aldehydes with H$_2$O$_2$, UHP, or sodium percarbonate in various solvents (Table 5.4) [71,72]. The epoxidation step appeared to proceed via a plausible combined iminium and enamine mechanism.

Hayashi and co-workers developed an asymmetric organocatalytic procedure for the enantioselective oxidation of α-substituted acroleins. In the presence of H$_2$O$_2$ (3 equivalents) and catalyst 17a (Figure 5.3), the acroleins were converted into cor-responding epoxides to 94% ee [73]. List and co-workers found a cinchona-derived amine catalyst 18 for the enantioselective epoxidation of α,β-substituted α,β-unsaturated aldehydes and α-substituted acroleins. Crucial to success was the com-bination of a chiral primary cinchona-based amine and a chiral phosphoric acid [74].

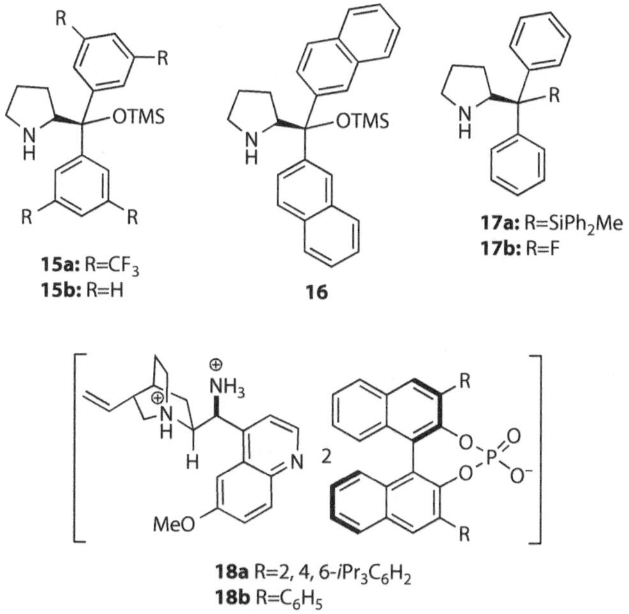

15a: R=CF$_3$
15b: R=H

16

17a: R=SiPh$_2$Me
17b: R=F

18a R=2, 4, 6-*i*Pr$_3$C$_6$H$_2$
18b R=C$_6$H$_5$

FIGURE 5.3 Phase transfer organic catalysts for enantioselective epoxidation of α,β-unsaturated aldehydes with H$_2$O$_2$.

Gilmour with co-workers prepared fluorinated catalyst **17b** that demonstrated good to excellent enantioselectivities in the epoxidations of β-substituted α,β-unsaturated aldehydes [75]. Kudo and Akagawa reported the first asymmetric oxidation of α,β-unsaturated aldehydes with H$_2$O$_2$ in the presence of a resin-supported polypeptide catalyst [76].

Organocatalytic asymmetric epoxidation of α,β-unsaturated aldehydes advanced greatly since 2000. Although the existing procedures are slow (hours or days are the usual reaction times) and typically require 10 to 20 mol% of a chiral organocatalyst, various versions of enantioselective epoxidations of α,β-unsaturated aldehydes have been exploited at laboratory scale to develop tandem or complex synthetic procedures such as tandem organocatalytic asymmetric synthesis of 1,2,3-triols [77], tandem epoxidation-Wittig and epoxidation-Mannich reaction sequences [72], enantioselective syntheses of *trans*-2,3-dihydroxyaldehydes and their derivatives, [78] β-hydroxy esters [79,80], electron-poor 2-hydroxyalkyl- and 2-aminoalkyl furanes [81], and α,β-epoxy esters [82].

MISCELLANEOUS OXIDATIONS

This section surveys other types of organocatalyzed asymmetric oxidations with H$_2$O$_2$ and O$_2$, in particular sulfoxidations, Baeyer–Villiger oxidations, and α-hydroxylations of carbonyl compounds.

TABLE 5.4

Asymmetric Organocatalytic Epoxidation of α,β-Unsaturated Aldehydes with H_2O_2 and Its Derivatives

N	Substrate	Catalyst	Oxidant	Additive	Epoxide (%)	ee (%)	Ref.
1		15a	H_2O_2	—	80	96	[69]
		15b	UHP	—	96	98	[72]
		16	H_2O_2	—	86	98	[72]
		17b	H_2O_2	—	92	96	[75]
2	EtO$_2$C	15b	H_2O_2	—	>90^a	98	[71]
3	C$_4$H$_9$	15b	$Na_2CO_3 \cdot 1.5H_2O_2$	—	70	94	[72]
4	C$_7$H$_{16}$	17a	H_2O_2	—	72	94	[73]
5	Ph	17a	H_2O_2	—	65	86	[73]
		18a	H_2O_2	—	78	98	[74]
6	C$_3$H$_3$, Et	18b	H_2O_2	—	64	98	[74]
7		17b	H_2O_2	—	68	97	[75]

^{a} Conversion reported.

SULFOXIDATIONS

In 1988, Shinkai et al. reported a chiral flavinium salt 19–H_2O_2 (Figure 5.4) catalyst system capable of conducting the asymmetric oxidation of aryl methyl sulfides to sulfoxides in aqueous methanolic solutions [83]. The planar–chiral organocatalyst performed up to eight catalytic turnovers, yielding sulfoxides in up to 65.4% *ee* (for methyl tolyl sulfide; Table 5.5). The authors suggested the structure of possible active species—a chiral flavinium hydroperoxide [83].

A structurally related amine 20 afforded methyl 2-naphthyl sulfoxide in 94% yield and 72% *ee* [84]. Much later, Cibulka and co-workers studied the oxidations of various alkyl aryl sulfides with H_2O_2 in the presence of chiral flavinium salts 21. Somewhat higher catalytic efficiencies and lower enantioselectivities (4 to 54% *ee*) were reported in water–methanol solutions [85]. Subsequently, they synthesized flavin-β-cyclodextrin conjugates of type 22 that demonstrated higher activities,

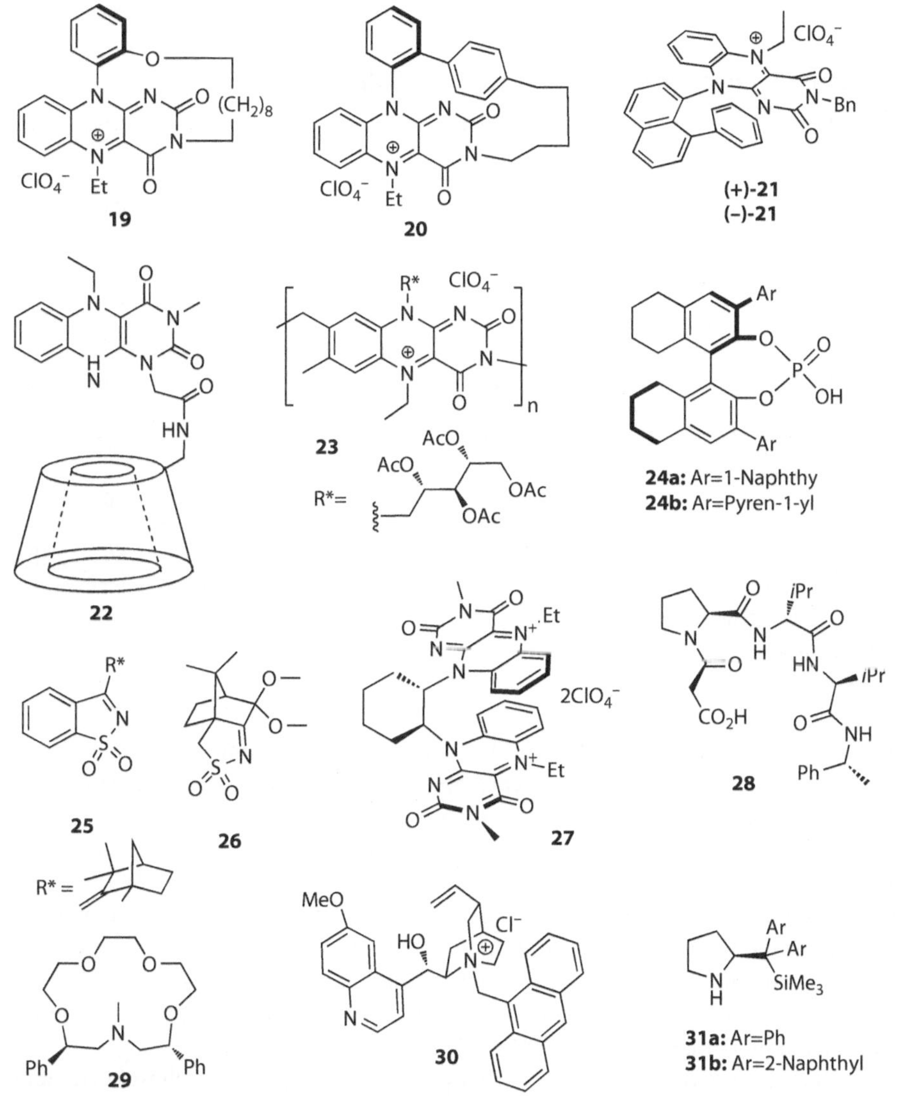

FIGURE 5.4 Organocatalysts used for miscellaneous asymmetric oxidations.

efficiencies, and enantioselectivities. Methyl *p*-tolyl sulfides were oxidized with 97% conversion and 80% *ee* within 10 min using only 1 mol% of catalyst [86]. It is noteworthy that the reactions proceeded in neat aqueous (buffered) media, and the catalyst loads could be further reduced to only 0.2 mol% (to achieve the maximum turnover of 395).

In a subsequent study, the authors reported even higher enantioselectivity (91% *ee*) in the oxidation of *t*-butyl methyl sulfide [87]. Yashima and co-workers prepared an optically active polymer **23** consisting of riboflavin units that catalyzed the asymmetric organocatalytic oxidation of sulfides with H_2O_2 with up to 60% *ee* [88]. This

TABLE 5.5
Asymmetric Organocatalytic Oxidation of Sulfides with H_2O_2

N	Substrate	Catalyst	Oxidant	Additive	Sulfoxide (%)	ee (%)	Ref.
1		19	H_2O_2	—	a	65	[83]
		(-)-21	H_2O_2	—	b	42	[85]
		22	H_2O_2	—	97c	80	[86]
		23	H_2O_2	—	39	60	[88]
		24a	H_2O_2	—	89	80	[89]
2		20	H_2O_2	—	94	72	[84]
		(-)-21	H_2O_2	—	b	54	[85]
3		(-)-21	H_2O_2	—	b	36	[85]
	MeO	22	H_2O_2	—	97c	72	[86]
		24a	H_2O_2	—	92	70	[89]
4		26	H_2O_2	—	100	86	[90]

a Eight catalytic turnovers reported.
b Not specified.
c Conversion reported.

enantioselectivity was much higher than that of a model monomeric counterpart (30% *ee*), thus indicating that the supramolecular structure of the catalyst contributed significantly to the observed oxidation enantioselectivity.

Tao and Wang with co-workers reported enantioselective sulfoxidations with H_2O_2 in the presence of chiral binol-derived phosphoric acids of type **24** [89]. With the catalyst of choice, **24a** (10 mol%), a series of aryl methyl sulfides were oxidized to the corresponding sulfoxides with moderate to high yields (35 to 92%) and up to 82% *ee* within 48 to 115 hr at −40°C. The oxidation of 2-aryl-substituted 1,3-dithianes showed a good monooxygenation selectivity (with preferential *trans*-oxide formation) and moderate *ees* (60 to 70%) [89].

To date, no organocatalysts other than flavin derivatives and phosphoric acids have proven capable of conducting sulfoxidations with H_2O_2 in a stereoselective fashion. Page et al. reported that camphor-derived *N*-sulfonyl imines of types **25** and **26** could mediate sulfoxidations with H_2O_2 in the presence of 4 equivalents of DBU. However, good enantioselectivities (up to 98% *ee*) were only achieved when the sulfonyl imines were employed in stoichiometric amounts [90,91].

Baeyer–Villiger Oxidations

Several examples of asymmetric organocatalytic Baeyer–Villiger oxidations have been reported. The first example of such a process was reported by Murahashi who demonstrated that planar–chiral bisflavinium perchlorate **27** (Figure 5.4) catalyzed the asymmetric Baeyer–Villiger reaction of 3-aryl substituted cyclobutanones with hydrogen peroxide (1.5 equivalents) to generate optically active γ-lactones in up to 74% *ee* with moderate to good yields ranging from 17 to 81% (Table 5.6) [92].

A substantial disadvantage of the oxidation protocol was the low reactivity. At 10 mol% catalyst loads, the reaction required 6 days at –30°C. Ding and co-workers later discovered that chiral phosphoric acids of type **24** could catalyze the Baeyer–Villiger oxidation of 3-substituted cyclobutanones with hydrogen peroxide [93,94]. With the catalyst of choice, **24b** (Ar=pyren-1-yl), the process featured high yields (65 to 99%) and moderate to high *ee*s (55 to 93%). With 10 mol% of catalyst, the reaction took 18 to 80 hr at –40°C to proceed to quantitative conversion [93].

Chiral phosphoric acid catalysts of type **24** were applied successfully to the enantioselective oxidation of several bicyclic and tricyclic cyclobutanones to achieve excellent yields and moderate to good *ee*s (for the minor "abnormal" regioisomer, up to 99% in a few cases) [95]. The mechanism of the catalytic Baeyer–Villiger transformation was examined via kinetic and computational studies [96].

Peris and Miller used aspartate-derived oligopeptide **28** as the chiral acid catalyst (25 mol%) for Baeyer–Villiger oxidative desymmetrization of cyclic ketones. The catalytic protocol required 25 mol% of catalyst, 12.5 equivalents of H_2O_2, 25 mol% of *N,N*-dimethyl-4-aminopyridine, and 10 equivalents of diisopropylcarbodiimide. In effect, two cyclic ketones yielded the corresponding lactones in 74 and 29% yields and with enantiomeric excesses of 30 and 42% *ee*, respectively [97].

TABLE 5.6
Asymmetric Organocatalytic Baeyer–Villiger Oxidations with H_2O_2

N	Substrate	Catalyst	Oxidant	Additive	Sulfoxide (%)	ee (%)	Ref.
1		27	H_2O_2	NaOAc	53	62	[92]
		24b	H_2O_2	—	99	93	[93]
2		27	H_2O_2	NaOAc	28	68	[92]
		24b	H_2O_2	—	99	83	[93]
3		27	H_2O_2	NaOAc	17	74	[92]
		24b	H_2O_2	BTCN	99	86	[94]

Note: BTCN = benzene-1,2,4,5-tetracarbonitrile.

α-Hydroxylations of Carbonyl Compounds

Asymmetric α-hydroxylation of carbonyl compounds is a challenging transformation because optically active α-hydroxy carbonyl compounds are used as starting materials and intermediates for the syntheses of natural products [98]. In 1988, Shioiri with co-workers reported the first organocatalytic method for enantioselective hydroxylation of prochiral ketones with molecular oxygen. Using catalyst **7e** (5 mol%) and aqueous NaOH and P(OEt)$_3$ (for the in situ reduction of the labile hydroperoxide intermediate), a series of 2-alkyl tetralones and indanones were oxidized with high yields (55 to 98%) and moderate to good enantioselectivities (27 to 79% *ee*) [98].

	Yield	*ee*
R=Me:	95%	70%
Et:	98%	72%
*i*Pr:	59%	77%

Dehmlow et al. screened a series of structural analogues of catalysts **7** and **8** that demonstrated low to moderate enantioselectivities (59% *ee* in one case) in α-hydroxylation of 2-ethyl tetralone with molecular oxygen [60]. Brusse with co-workers exploited chiral crown ether **29** as a catalyst for the same process under similar reaction conditions and demonstrated enantioselectivities similar to those of catalyst **7e** [99].

Itoh and co-workers developed an asymmetric aerobic hydroxylation procedure of 3-substituted oxindoles, relying on phase transfer catalyst **30**. At 20 mol% catalyst load, the reaction proceeded in a reasonably short time (usually 2.5 to 24 hr) to afford the corresponding chiral hydroxyketone in high yields and moderate to excellent enantioselectivities (up to 93% *ee*) [100].

Córdova et al. proposed a novel synthetic route to chiral 1,2-diols. They reported that *L*-proline catalyzed the asymmetric incorporation of singlet oxygen $^1O_2$ (photo-generated in the presence of tetraphenylporphyrin) at the α-positions of aldehydes [101]. Subsequent in situ reduction of the resulting hydroperoxy aldehyde intermediate afforded the target optically active terminal diol. Triplet dioxygen $^3O_2$ did not afford the diols [101].

The authors later examined the catalytic properties of a series of protected diarylprolinols of type **31** and identified catalysts **31a,b** as the most stereoselective (at 10 to 20 mol% loads), yielding chiral diols with up to 98% *ee*, albeit with moderate yields (50 to 76%) [102]. In chloroform or methanol, the enantioselectivities were higher than in DMF.

A series of D- and L-amino acids were tested as catalysts in catalytic asymmetric α-hydroxylation of ketones under UV irradiation [103]. Valine and alanine catalyzed the transformation with higher enantioselectivity than proline. Curiously, the reaction did not require external reductants (like P(OEt)$_3$), affording α-hydroxy ketones with moderate to excellent yields and moderate enantioselectivities (up to 72% *ee* in one case; Table 5.7). Conversely, the treatment of the reaction mixture with NaBH$_4$ led to a mixture of *cis*- and *trans*-diols, with retention of optical purity of the *trans* product [103].

Interesting and promising examples of organocatalyzed α-hydroxylations with O$_2$ have been reported. Their major shortcomings are high catalyst loads (10 to 20 mol%) and in most cases a need for an external co-reductant.

In spite of the milestone success of organocatalyzed Juliá-Colonna epoxidation of chalcones that attracted a significant interest of key chemical producers like Bayer AG and Degussa AG [1,3,6], the search for and rational design of environmentally sustainable preparative asymmetric organocatalytic oxidations is in its infancy. Nevertheless, known organocatalysts capable of mediating asymmetric oxidations with hydrogen peroxide and dioxygen have already demonstrated great synthetic potential.

At the same time, most of them feature serious drawbacks, such as narrow substrate scope (e.g., electron-deficient olefins, mostly E-enones, simple alkyl aryl sulfides, or 3-substituted cyclobutanones), low catalytic efficiencies, and the need for external co-reductants for oxidations with O$_2$. Asymmetric organocatalytic oxidations may be

---

TABLE 5.7

Asymmetric Organocatalyzed α-Hydroxylation of Carbonyl Compounds with O$_2$

N	Substrate	Catalyst	Oxidant	Additive	Yield (%)	ee (%)	Ref.
1	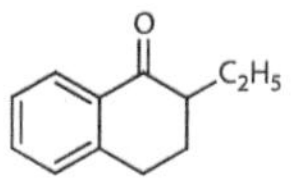 (ketone with C$_2$H$_5$)	7e	O$_2$	NaOH	98	72	[98]
		29	O$_2$	NaOH	95	67	[99]
		8 (R = Ph)	O$_2$	NaOH	60	50	[60]
2	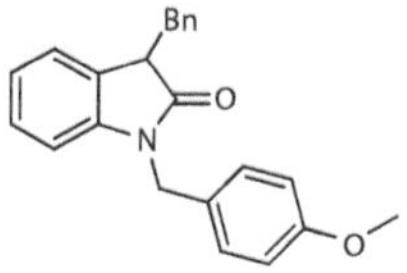 (oxindole with Bn)	30	air	NaOH	95	86	[100]
3	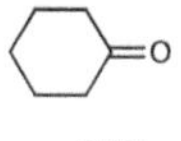 (cyclohexanone)	*L*-alanine	O$_2$	TPP	93	56	[103]
4	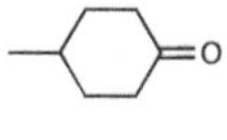 (methylcyclohexanone)	*L*-alanine	O$_2$	TPP	67	72	[103]

Note: TPP = tetraphenylporphyrin.

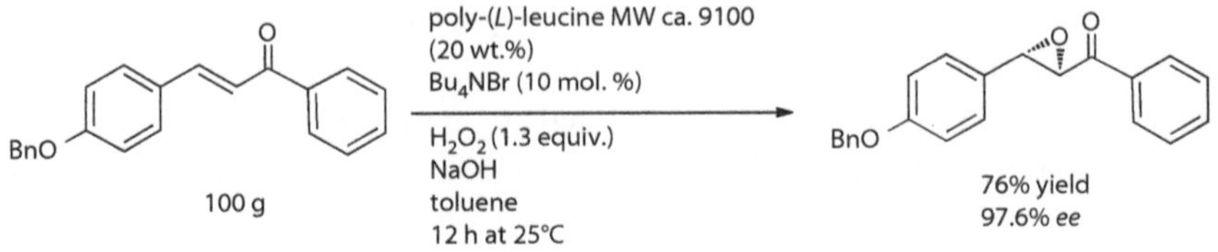

Hundred-gram scale Julia-Colonna enantioselective epoxidation of substituted chalcone [37].

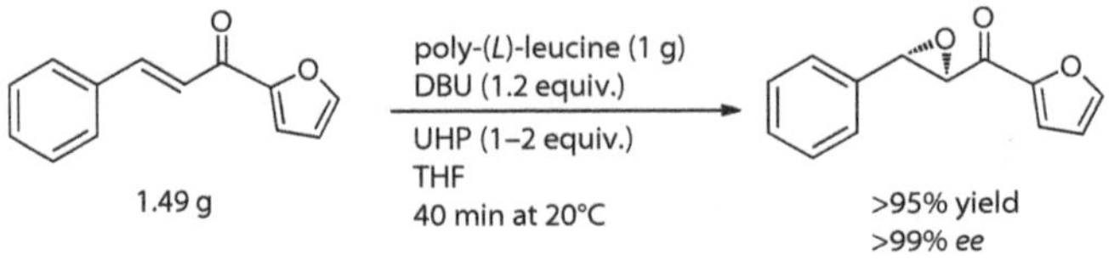

Enantioselective epoxidation of furyl styryl ketone–key step of the synthesis of naturally occurring styryl lactones (+)-goniofufurone, (+)-8-acetylgoniotriol, (+)-goniopypyrone [48].

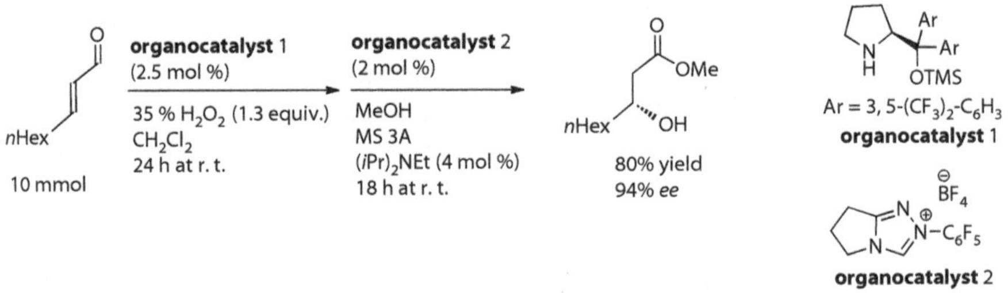

A one-pot reaction sequence for the synthesis of enantioenriched β-hydroxy esters [80]. First step: organocatalytic asymmetric epoxidation of an α, β-unsaturated aldehyde.

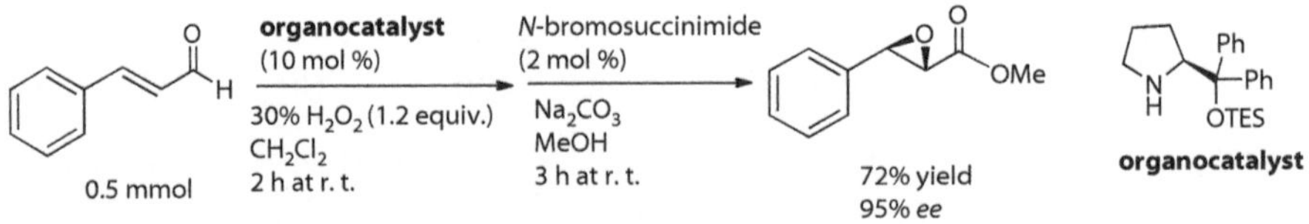

Organocatalytic synthesis of an α, β-epoxy ester, further used for the preparation of (−)-clausenamide [82].

FIGURE 5.5 Examples of preparative scale organocatalytic asymmetric transformations.

regarded as complementary to metal-mediated techniques since organic compounds may better catalyze processes that are difficult to perform than conventional metal-based catalysts, for example, the epoxidation of α,β-unsaturated aldehydes or trisubstituted olefins bearing carbamate functionalities [49] with H_2O_2 or α-hydroxylation of carbonyl compounds with O_2.

Most organocatalyzed asymmetric oxidation processes are still a long way from practical application. Further improvements are expected, mainly to broaden substrate scope, reduce oxidant consumption to stoichiometric or slightly higher levels, and increase catalyst efficiencies, scalabilities, and stereoselectivities.

REFERENCES

1. Berkessel, A. and Gröger, H. 2005. *Asymmetric Organocatalysis*. Weinheim: Wiley-VCH.
2. Dalko, P. I., Ed. 2007. *Enantioselective Organocatalysis*. Weinheim: Wiley-VCH.
3. Gröger, H. 2008. Asymmetric organocatalysis on a technical scale: current status and future challenges. *Ernst Schering Found. Symp. Proc.* 2: 141–158.
4. Wong, O. A. and Shi, Y. 2008. Organocatalytic oxidation: asymmetric epoxidation of olefins catalyzed by chiral ketones and iminium salts. *Chem. Rev.* 108: 3958–3987.
5. Kelly, D. R. and Roberts, S. M. 2006. Oligopeptides as catalysts for asymmetric epoxidation. *Biopoly. Peptide Sci.* 84: 74–89.
6. Weiss, K. M. and Tsogoeva, S. B. 2011. Enantioselective epoxidation of electron-deficient olefins: an organocatalytic approach. *Chem. Rec.* 11: 18–39.
7. Russo, A., De Fusco, C., and Lattanzi, A. 2012. Organocatalytic asymmetric oxidations with hydrogen peroxide and molecular oxygen. *ChemCatChem* 4: 901–916.
8. Porter, M. J. and Skidmore, J. 2000. Asymmetric epoxidation of electron-deficient olefins. *Chem. Commun.* 1215–1225.
9. Frohn, M. and Shi, Y. 2000. Chiral ketone-catalyzed asymmetric epoxidation of olefins. *Synthesis* 1979–2000.
10. Porter, M. J. and Skidmore, J. 2010. Asymmetric epoxidation of electron-deficient alkenes. In *Organic Reactions*, Vol. 74; Denmark, S. E., Ed. New York: John Wiley & Sons, pp. 426–472.
11. Diez, D., Nunez, M. G., Anton, A. B. et al. 2008. Asymmetric epoxidation of electron-deficient olefins. *Curr. Org. Synth.* 5: 186–216.
12. Shu, L. and Shi, Y. 1999. Asymmetric epoxidation using hydrogen peroxide (H_2O_2) as primary oxidant. *Tetrahedron Lett.* 40: 8721–8724.
13. Shu, L. and Shi, Y. 2001. An efficient ketone-catalyzed asymmetric epoxidation using hydrogen peroxide (H_2O_2) as primary oxidant. *Tetrahedron* 57: 5213–5218.
14. Burke, C. P., Shu, L., and Shi, Y. 2007. An *N*-aryl-substituted oxazolidinone-containing ketone-catalyzed asymmetric epoxidation with hydrogen peroxide as the primary oxidant. *J. Org. Chem.* 72: 6320–6323.
15. Zhang, A., Tang, J., Wang, X. et al. 2008. Chiral ketone- or chiral amine-catalyzed asymmetric epoxidation of *cis*-1-propenylphosphonic acid using hydrogen peroxide as oxidant. *J. Mol. Catal. A Chem.* 285: 68–71.
16. Page, P. C. B., Parker, P., Rassia, G. A. et al. 2008. Iminium salt-catalysed asymmetric epoxidation using hydrogen peroxide as stoichiometric oxidant. *Adv. Synth. Catal.* 350: 1867–1874.
17. Juliá, S., Masana, J., and Vega, J. C. 1980. Synthetic enzymes: highly stereoselective epoxidation of chalcone in a triphasic toluene–water–poly(*S*)-alanine system. *Angew. Chem. Int. Ed. Engl.* 19: 929–931.
18. Juliá, S., Masana, J., Rocas, J. et al. 1982. Synthetic enzymes 2. Catalytic asymmetric epoxidation by means of polyamino acids in a triphase system. *J. Chem. Soc. Perkin Trans.* 1317–1324.
19. Juliá, S., Masana, J., Rocas, J. et al. 1983. Synthetic enzymes 3. Highly stereoselective epoxidation of chalcones in a triphasic toluene–water–poly(*S*)-alanine. system. *Anal. Quim. Ser. C* 79: 102–104.

20. Colonna, S., Molinari, H., Banfi, S. et al. 1983. Synthetic enzymes 4. Highly enanti-oselective epoxidation by means of polyamino acids in a triphase system: influence of structural variations within the catalysts. *Tetrahedron* 39: 1635–1641.

21. Banfi, S., Colonna, S., Molinari, H. et al. 1984. Asymmetric epoxidation of electron-poor olefins 5. Influence on stereoselectivity of the structure of poly-α-amino acids used as catalysts. *Tetrahedron* 40: 5207–5211.

22. Kroutil, W., Mayon, P., Lasterra-Sánchez, M. E. et al. 1996. Unexpected asymmet-ric epoxidation reactions catalysed by polyleucine-based systems. *Chem. Commun.* 845–846.

23. Bentley, P. A., Bickley, J. F., Roberts, S. M. et al. 2001. Asymmetric epoxidation of a geminally disubstituted and some trisubstituted enones catalysed by poly-*L*-leucine. *Tetrahedron Lett.* 42: 3741–3743.

24. Bentley, P. A., Bergeron, S., Cappi, M. W. et al. 1997. Asymmetric epoxidation of enones employing polymeric α-amino acids in non-aqueous media. *Chem. Commun.* 739–740.

25. Geller, T. and Roberts, S. M. 1999. A new procedure for the Juliá-Colonna stereoselec-tive epoxidation reaction under non-aqueous conditions: development of a catalyst com-prising polyamino acid on silica (PaaSiCat). *J. Chem. Soc. Perkin Trans.* 1397–1398.

26. Cappi, M. W., Chen, W. P., Flood, R. W. et al. 1998. New procedures for the Juliá-Colonna asymmetric epoxidation: synthesis of (+)-clausenamide. *Chem. Commun.* 1159–1160.

27. Allen, J. V., Drauz, K. H., Flood, R. W. et al. 1999. Polyamino acid-catalysed asymmet-ric epoxidation: sodium percarbonate as a source of base and oxidant. *Tetrahedron Lett.* 40: 5417–5420.

28. Allen, J. V., Bergeron, S., Griffith, M. J. et al. 1998. Juliá-Colonna asymmetric epoxidation reactions under non-aqueous conditions: rapid, highly regio- and stereo-selective transfor-mations using a cheap, recyclable catalyst. *J. Chem. Soc. Perkin Trans.* 3171–3180.

29. Carde, L., Davies, H., Geller, T. P. et al. 1999. PaaSicats: powerful catalysts for asym-metric epoxidation of enones: novel syntheses of α-arylpropanoic acids including (*S*)-fenoprofen. *Tetrahedron Lett.* 40: 5421–5424.

30. Itsuno, S., Sakakura, M., and Ito, K. 1990. Polymer-supported poly(amino acids) as new asymmetric epoxidation catalyst of α,β-unsaturated ketones. *J. Org. Chem.* 55: 6047–6049.

31. Dhanda, A., Drauz, K. H., Geller, T. P. et al. 2000. PaaSiCats: novel polyamino acid catalysts. *Chirality* 12: 313–317.

32. Yi, H., Zou, G., Li, Q. et al. 2005. Asymmetric epoxidation of α,β-unsaturated ketones catalyzed by silica-grafted poly-(*L*)-leucine catalysts. *Tetrahedron Lett.* 46: 5665–5668.

33. Berkessel, A., Koch, B., Toniolo, C. et al. 2006. Asymmetric enone epoxidation by short solid-phase bound peptides: further evidence for catalyst helicity and catalytic activity of individual peptide strands. *Biopolymers (Peptide Science)* 84: 90–96.

34. Geller, T., Gerlach, A., Krüger, C. M. et al. 2004. Novel conditions for the Juliá-Colonna epoxidation reaction providing efficient access to chiral, non-racemic epox-ides. *Tetrahedron Lett.* 45: 5065–5067.

35. Geller, T., Krüger, C. M., and Militzer, H. C. 2004. Scoping triphasic PTC conditions for the Juliá-Colonna epoxidation reaction. *Tetrahedron Lett.* 45: 5069–5071.

36. Geller, T., Gerlach, A., Kruger, C. M. et al. 2006. The Juliá-Colonna epoxidation: access to chiral, non-racemic epoxides. *J. Mol. Catal. A Chem.* 251: 71–77.

37. Gerlach, A. and Geller, T. 2004. Scale-up studies for the asymmetric Juliá-Colonna epoxidation reaction. *Adv. Synth. Catal.* 346: 1247–1249.

38. Qiu, W., He, L., Chen, Q. et al. 2009. Imidazolium-modified poly(*L*-leucine) catalyst: an efficient and recoverable catalyst for Juliá-Colonna reactions. *Tetrahedron Lett.* 50: 5225–5227.

39. Bentley, P. A., Cappi, M. W., Flood, R. W. et al. 1998. Toward a mechanistic insight into the Juliá-Colonna asymmetric epoxidation of α,β-unsaturated ketones using discrete lengths of polyleucine. *Tetrahedron Lett.* 39: 9297–9300.

40. Bentley, P. A., Flood, R. W., Roberts, S. M. et al. 2001. The effect of the primary structure of the polypeptide catalyst on the enantioselectivity of the Juliá-Colonna asymmetric epoxidation of enones. *Chem. Commun.* 1616–1617.

41. Berkessel, A., Gasch, N., Glaubitz, K. et al. 2001. Highly enantioselective enone epoxidation catalyzed by short solid phase-bound peptides: dominant role of peptide helicity. *Org. Lett.* 3: 3839–3842.

42. Weyer, A., Díaz, D., Nierth, A. et al. 2012. The *L*-leu hexamer, a short and highly enantioselective peptide catalyst for the Juliá-Colonna epoxidation: stabilization of a helical conformation in DMSO. *ChemCatChem* 4: 337–340.

43. Takagi, R., Shiraki, A., Manabe, T. et al. 2000. The Juliá-Colonna asymmetric epoxidation reaction catalyzed by soluble oligo-*L*-leucines containing an α-aminoisobutyric acid residue: importance of helical structure of the catalyst on asymmetric induction. *Chem. Lett.* 366–367.

44. Flood, R. W., Geller, T. P., Petty, S. A. et al. 2001. Efficient asymmetric epoxidation of α,β-unsaturated ketones using a soluble triblock polyethylene glycol–polyamino acid catalyst. *Org. Lett.* 3: 683–686.

45. Tsogoeva, S. B., Woltinger, J., Jost, C. et al. 2002. Juliá-Colonna asymmetric epoxidation in a continuously operated Chemzyme membrane reactor. *Synlett.* 707–709.

46. Kelly, D. R., Bui, T. T. T., Caroff, E. et al. 2004. Structure and catalytic activity of some soluble polyethylene glycol–peptide conjugates. *Tetrahedron Lett.* 45: 3885–3888.

47. Adger, B. M., Barkley, J. V., Bergeron, S. et al. 1997. Improved procedure for Juliá-Colonna asymmetric epoxidation of α,β-unsaturated ketones: total synthesis of diltiazem and Taxol™ side-chain. *J. Chem. Soc. Perkin Trans.* 3501–3507.

48. Chen, W. P. and Roberts, S. M. 1999. Juliá-Colonna asymmetric epoxidation of furyl styryl ketone as a route to intermediates to naturally occurring styryl lactones. *J. Chem. Soc. Perkin Trans.* 103–106.

49. Peris, G., Jakobsche, C. E., and Miller, S. J. 2007. Aspartate-catalyzed asymmetric epoxidation reactions. *J. Am. Chem. Soc.* 129: 8710–8711.

50. Jakobsche, C. E., Peris, G., and Miller, S. J. 2008. Functional analysis of an aspartate-based epoxidation catalyst with amide-to-alkene peptidomimetic catalyst analogues. *Angew. Chem. Int. Ed.* 47: 6707–6711.

51. Kolundzic, F., Noshi, M. N., Tjandra, M. et al. 2011. Chemoselective and enantioselective oxidation of indoles employing aspartyl peptide catalysts. *J. Am. Chem. Soc.* 133: 9104–9111.

52. Helder, R., Hummelen, J. C., Laane, R. W. et al. 1976. Catalytic asymmetric induction in oxidation reactions: synthesis of optically active epoxides. *Tetrahedron Lett.* 17: 1831–1834.

53. Wynberg, H. and Greijdanus, B. 1978. Solvent effects in homogeneous catalysis. *J. Chem. Soc. Chem. Commun.* 427–428.

54. Marsman, B. and Wynberg, H. 1979. Absolute configuration of chalcone epoxide. chemical correlation. *J. Org. Chem.* 44: 2312–2314.

55. Hummelen, J. C. and Wynberg, H. 1978. Alkaloid assisted asymmetric synthesis IV. Additional routes to chiral epoxides. *Tetrahedron Lett.* 19: 1089–1092.

56. Wynberg, H., and Marsman, B. 1980. Synthesis of optically active 2,3-epoxy cyclohexanone and the determination of its absolute configuration. *J. Org. Chem.* 45: 158–161.

57. Pluim, H. and Wynberg, H. 1980. Catalytic asymmetric induction in oxidation reactions: synthesis of optically active epoxynaphthoquinones. *J. Org. Chem.* 45: 2498–2502.

58. Arai, S., Tsuge H., and Shioiri, T. 1998. Asymmetric epoxidation of α,β-unsaturated ketones under phase transfer-catalyzed conditions. *Tetrahedron Lett.* 39: 7563–7566.

59. Arai, S., Tsuge, H., Oku, M. et al. 2002. Catalytic asymmetric epoxidation of enones under phase transfer-catalyzed conditions. *Tetrahedron* 58: 1623–1630.

60. Dehmlow, E. V., Düttmann, D., Neumann, B. et al. 2002. Monodeaza cinchona alkaloid derivatives: synthesis and preliminary applications as phase transfer catalysts. *Eur. J. Org. Chem.* 2087–2093.

61. Liu, X. D., Bai, X. L., Qiu, X. P. et al. 2005. Asymmetric phase transfer-mediated epoxidation of α,β-enones using dendritic catalysts derived from cinchona alkaloids. *Chin. Chem. Lett.* 16: 975–978.

62. Jew, S. S., Lee, J. H., Jeong, B.S. et al. 2005. Highly enantioselective epoxidation of 2,4-diarylenones using dimeric cinchona phase transfer catalysts: enhancement of enantioselectivity by surfactants. *Angew. Chem. Int. Ed.* 44: 1383–1385.

63. Hori, K., Tamura, M., Tani, K. et al. 2006. Asymmetric epoxidation catalyzed by novel azacrown ether-type chiral quaternary ammonium salts under phase transfer catalytic conditions. *Tetrahedron Lett.* 47: 3115–3118.

64. Reisinger, C. M., Wang X., and List, B. 2008. Catalytic asymmetric hydroperoxidation of α,β-unsaturated ketones: approach to enantiopure peroxyhemiketals, epoxides, and aldols. *Angew. Chem. Int. Ed.* 47: 8112–8115.

65. Wang X., Reisinger, C. M., and List, B. 2008. Catalytic asymmetric epoxidation of cyclic enones. *J. Am. Chem. Soc.* 130: 6070–6071.

66. Lee, A., Reisinger, C., and List, B. 2012. Catalytic asymmetric epoxidation of 2-cyclopentenones. *Adv. Synth. Catal.* 354: 1701–1706.

67. Tanaka, S. and Nagasawa, K. 2009. Guanidine–urea bifunctional organocatalyst for asymmetric epoxidation of 1,3-diarylenones with hydrogen peroxide. *Synlett.* 667–670.

68. Terada, M. and Nakano, M. 2008. Asymmetric epoxidation of α,β-unsaturated ketones with hydrogen peroxide catalyzed by axially chiral guanidine base. *Heterocycles* 76: 1049–1055.

69. Marigo, M., Franzén, J., Poulsen, T. B. et al. 2005. Asymmetric organocatalytic epoxidation of α,β-unsaturated aldehydes with hydrogen peroxide. *J. Am. Chem. Soc.* 127: 6964–6965.

70. Zhuang, W., Marigo, M., and Jørgensen, K. A. 2005. Organocatalytic asymmetric epoxidation reactions in water-alcohol solutions. *Org. Biomol. Chem.* 3: 3284–3289.

71. Sundén, H., Ibrahem, I., and Córdova, A. 2006. Direct organocatalytic asymmetric epoxidation of α,β-unsaturated aldehydes. *Tetrahedron Lett.* 47: 99–103.

72. Zhao, G. L., Ibrahem, I., Sundén, H. et al. 2007. Amine-catalyzed asymmetric epoxidation of α,β-unsaturated aldehydes. *Adv. Synth. Catal.* 349: 1210–1224.

73. Bondzic, B. P., Urushima, T., Ishikawa, H. et al. 2010. Asymmetric epoxidation of α-substituted acroleins catalyzed by diphenylprolinol silyl ether. *Org. Lett.* 12: 5434–5437.

74. Lifchits, O., Reisinger, C. M., and List, B. 2010. Catalytic asymmetric epoxidation of α-branched enals. *J. Am. Chem. Soc.* 132: 10227–10229.

75. Sparr, C., Schweizer, W. B., Senn, H. M. et al. 2009. The fluorine–iminium ion gauche effect: proof of principle and application to asymmetric organocatalysis. *Angew. Chem. Int. Ed.* 48: 3065–3068.

76. Akagawa, K. and Kudo, K. 2011. Asymmetric epoxidation of α,β-unsaturated aldehydes in aqueous media catalyzed by resin-supported peptide-containing unnatural amino acids. *Adv. Synth. Catal.* 353: 843–847.

77. Zhao, G.-L., Dziedzic, P., Ibrahem, I. et al. 2006. Organocatalytic asymmetric synthesis of 1,2,3-*prim,sec,sec*-triols. *Synlett.* 3521–3524.

78. Albrecht, L., Jiang, H., Dickmeiss, G. et al. 2010. Asymmetric formal *trans*-dihydroxylation and *trans*-aminohydroxylation of α,β-unsaturated aldehydes via an organocatalytic reaction cascade. *J. Am. Chem. Soc.* 132: 9188–9196.

79. Zhao, G. L. and Córdova, A. 2007. A one-pot combination of amine and heterocyclic carbene catalysis: direct asymmetric synthesis of β-hydroxy and β-malonate esters from α,β-unsaturated aldehydes. *Tetrahedron Lett.* 48: 5976–5980.

80. Jiang, H., Gschwend, B., Albrecht, L. et al. 2010. Organocatalytic preparation of simple β-hydroxy and β-amino esters: low catalyst loadings and gram-scale synthesis. *Org. Lett.* 12: 5052–5055.

81. Albrecht, L., Ransborg, L. K., Gschwend, B. et al. 2010. An organocatalytic approach to 2-hydroxyalkyl- and 2-aminoalkyl furanes. *J. Am. Chem. Soc.* 132: 17886–17893.

82. Xuan, Y., Lin, H. S., and Yan, M. 2013. Highly efficient asymmetric synthesis of α,β-epoxy esters *via* one-pot organocatalytic epoxidation and oxidative esterification. *Org. Biomol. Chem.* 11: 1815–1817.

83. Shinkai, S., Yamaguchi, T., Manabe, O. et al. 1988. Enantioselective oxidation of sulfides with chiral 4a-hydroperoxyflavin. *J. Chem. Soc. Chem. Commun.* 1399–1401.

84. Murahashi, S. I. 1995. Synthetic aspects of metal-catalyzed oxidations of amines and related reactions. *Angew. Chem. Int. Ed. Engl.* 34: 2443–2465.

85. Jurok, R., Cibulka, R., Dvořáková, H. et al. 2010. Planar chiral flavinium salts: prospective catalysts for enantioselective sulfoxidation reactions. *Eur. J. Org. Chem.* 5217–5224.

86. Mojr, V., Herzig, V., Buděšínsky, M. et al. 2010. Flavin–cyclodextrin conjugates as catalysts of enantioselective sulfoxidations with hydrogen peroxide in aqueous media. *Chem. Commun.* 46: 7599–7601.

87. Mojr, V., Buděšínsky, M., Cibulka, R. et al. 2011. Alloxazine–cyclodextrin conjugates for organocatalytic enantioselective sulfoxidations. *Org. Biomol. Chem.* 9: 7318–7326.

88. Iida, H., Iwahana, S., Mizoguchi, T. et al. 2012. Main-chain optically active riboflavin polymer for asymmetric catalysis and its vapochromic behavior. *J. Am. Chem. Soc.* 134: 15103–15113.

89. Liu, Z. M., Zhao, H., Li, M. Q. et al. 2012. Chiral phosphoric acid-catalyzed asymmetric oxidation of aryl alkyl sulfides and aldehyde-derived 1,3-dithianes using aqueous hydrogen peroxide as the terminal oxidant. *Adv. Synth. Catal.* 354: 1012–1022.

90. Page, P. C. B., Heer, J. P., Bethell, D. et al. 1999. Asymmetric sulfur oxidation mediated by camphor sulfonylimines. *Phosphorus Sulfur Silicon* 153: 247–258.

91. Bethell, D., Page, P. C. B., and Vahedi, H. 2000. Catalytic asymmetric oxidation of sulfides to sulfoxides mediated by chiral 3-substituted-1,2-benzisothiazole 1,1-dioxides. *J. Org. Chem.* 65: 6756–6760.

92. Murahashi, S. I., Ono, S., and Imada, Y. 2002. Asymmetric Baeyer–Villiger reaction with hydrogen peroxide catalyzed by a novel planar–chiral bisflavin. *Angew. Chem. Int. Ed.* 41: 2366–2368.

93. Xu, S., Wang, Z., Zhang, X. et al. 2008. Chiral Brønsted acid-catalyzed asymmetric Baeyer–Villiger reaction of 3-substituted cyclobutanones using aqueous H₂O₂. *Angew. Chem. Int. Ed.* 47: 2840–2843.

94. Xu, S., Wang, Z., Zhang, X. et al. 2010. Charge transfer effect on chiral phosphoric acid-catalyzed asymmetric Baeyer–Villiger oxidation of 3-substituted cyclobutanones using 30% aqueous H₂O₂ as the oxidant. *Chin. J. Chem.* 28: 1731–1735.

95. Xu, S., Wang, Z., Zhang, X. et al. 2011. Asymmetric Baeyer–Villiger oxidation of 2,3- and 2,3,4-substituted cyclobutanones catalyzed by chiral phosphoric acids with aqueous H₂O₂ as the oxidant. *Eur. J. Org. Chem.* 110–116.

96. Xu, S., Wang, Z., Li, Y. et al. 2010. Mechanistic investigation of chiral phosphoric acid-catalyzed asymmetric Baeyer–Villiger reaction of 3-substituted cyclobutanones with H₂O₂ as the oxidant. *Chem. Eur. J.* 16: 3021–3035.

97. Peris, G. and Miller, S. J. 2008. A non-enzymatic acid–peracid catalytic cycle for the Baeyer–Villiger oxidation. *Org. Lett.* 10: 3049–3052.

98. Masui, M., Ando, A., and Shioiri, T. 1988. New methods and reagents in organic synthesis: asymmetric synthesis of α-hydroxy ketones using chiral phase transfer catalysts. *Tetrahedron Lett.* 29: 2835–2838.

99. de Vries, E. F. J., Ploeg, L., Colao, M. et al. 1995. Enantioselective oxidation of aromatic ketones by molecular oxygen catalyzed by chiral monoaza-crown ethers. *Tetrahedron Asymmetry* 6: 1123–1132.

100. Sano, D., Nagata, K., and Itoh, T. 2008. Catalytic asymmetric hydroxylation of oxindoles by molecular oxygen using a phase transfer catalyst. *Org. Lett.* 10: 1593–1595.

101. Córdova, A., Sundén, H., Engqvist, M. et al. 2004. Direct amino acid-catalyzed asymmetric incorporation of molecular oxygen to organic compounds. *J. Am. Chem. Soc.* 126: 8914–8915.

102. Ibrahem, I., Zhao, G. L., Sundén, H. et al. 2006. A route to 1,2-diols by enantioselective organocatalytic α-oxidation with molecular oxygen. *Tetrahedron Lett.* 47: 4659–4663.

103. Sundén, H., Engqvist, M., Casas, J. et al. 2004. Direct amino acid-catalyzed asymmetric α-oxidation of ketones with molecular oxygen. *Angew. Chem. Int. Ed.* 43: 6532–6535.

6 Fe- and Mn-Based Synthetic Models of Non-Heme Oygenases

Stereospecific C–H Oxidations

This chapter covers catalytic processes of stereospecific oxidations of C–H functional groups. Strictly speaking, stereospecific processes are not asymmetric in the traditional sense since no new elements of chirality are created during catalytic transformation [1]. On the other hand, according to IUPAC definitions, stereospecific processes are necessarily stereoselective [1]. Moreover, multi-stage chiral synthesis (synthesis of chiral compounds) is always a combination of stereoselective chemical transformations (creating new asymmetric center(s)) and stereospecific ones (retaining the stereoconfiguration of already existing asymmetric centers).

From a practical synthetic perspective, the latter type of transformation is highly important. Interestingly, stereospecific chemical transformations may be catalyzed by the same catalysts as enantioselective ones. This is particularly true for catalyst systems for asymmetric epoxidation of olefins with H_2O_2, based on aminopyridine–iron and –manganese complexes (Chapter 2): those catalysts are also capable of conducting selective hydroxylation of C–H groups in a stereospecific fashion.

Nature successfully solved the problem of selective oxidation of non-activated C–H groups under mild conditions over the course of evolution, relying on the use of metalloenzymes (oxygenases) [2–5]. However, artificial catalyst systems demonstrating comparable levels of activity, selectivity, and efficiency are still rare and their design is highly challenging [6]. The catalysts discussed in this chapter were developed in the framework of the so-called biomimetic approach that involves modeling the catalytic activities of natural metalloenzymes by synthetic metallocomplexes [7].

Within this paradigm, modeling of non-heme monooxygenases is of particular interest, since they catalyze a broad range of oxidative transformations and are regarded as ideal platforms for catalysis [5]. Non-heme complexes of iron and manganese stand apart due to their availability and the high chemo- and stereoselectivities demonstrated in a variety of catalyzed oxidation processes [8]. Several specialized review papers have covered biomimetic oxidations and catalyzed C–H transformations [2–23]. Most of them focus on either the structures and spectroscopy or mechanistic features of the oxygenation process. The scopes and synthetic powers of such catalyst systems have not been surveyed systematically. This chapter discusses non-heme iron- and manganese-based systems capable of catalyzing oxidations of alkane C–H groups with H_2O_2; the issue of oxidation stereospecificity is given attention where appropriate.

IRON SYSTEMS

The first non-heme iron catalyst systems capable of conducting alkane hydroxylations with hydrogen peroxide appeared in the early 1990s. Sawyer with co-workers reported the oxidation of cyclohexane to cyclohexanone with H_2O_2 catalyzed by [Fe(pa)$_2$], [(pa)$_2$FeOFe(pa)$_2$], or [(dpa)FeOFe(dpa)] (pa (**1**) = pyridine-2-carboxylate; dpa (**2**) = pyridine-2,6-dicarboxylate; see Figure 6.1 [24]. The catalytic selectivity altered dramatically in different solvents. While the cyclohexanol:cyclohexanone ratio was 3:2 in acetonitrile, it changed to 6:94 in pyridine–acetic acid, accompanied by an eight-fold increase in efficiency [24].

Ketonization of other substrates was probed but in no cases did the turnover numbers exceed 23 [24]. Since the first milestone works, many iron complexes with related catalytic properties were reported (most were examined in acetonitrile solutions). Detailed accounts of those catalyst systems can be found in reviews [8,10,11,19]. In very few cases, however, the issue of hydroxylation stereospecificity was discussed.

Along with metal-based oxidants, free radicals generated from hydrogen peroxide decomposition may also be involved in hydrocarbon oxidation processes [7–10] and may affect the oxidation chemoselectivity and stereospecificity. To distinguish between the metal-based oxidation and free-radical-driven processes, a few test reactions are commonly used. For example, for the oxidation of cyclohexane, an alcohol:ketone (A:K) ratio close to 1.0 indicates diffusion-controlled free radical oxidation proceeding via alkylperoxyl radicals [8,9].

For metal-based oxygen rebound mechanisms, alcohol is the major oxidation product. This logic is correct when a high excess of substrate is used; otherwise, significant amounts of ketone may be formed via further oxidation of the alcohol [8–10].

Similarly, retention of absolute configuration (RC) in the course of hydroxylation of *cis*- and *trans*-1,2-dimethylcyclohexanes (complete or partial; see scheme), is evidence of metal-based oxidants. Oxidations via free radical pathways yield a mixture of epimeric alcohols [9,25,26]. We will refer to the A:K and RC parameters later in the text.

| *cis*-1,2-DMCH | *cis*-alcohol | *trans*-alcohol | 2-ketone | 3-ketone |
| | (major) | (minor) | (minor) | |

In 1995, Shteinman with co-workers examined the effects of electron-donating and -withdrawing substituents on the oxidation of alkanes with H_2O_2 in the presence of dinuclear catalysts [Fe$_2$OL$_2$(H$_2$O)$_2$](ClO$_4$)$_4$, where L = **3** or **4** [27,28]. In the course of epoxidation of *trans*-1,2-dimethylcyclohexane, partial retention of configuration

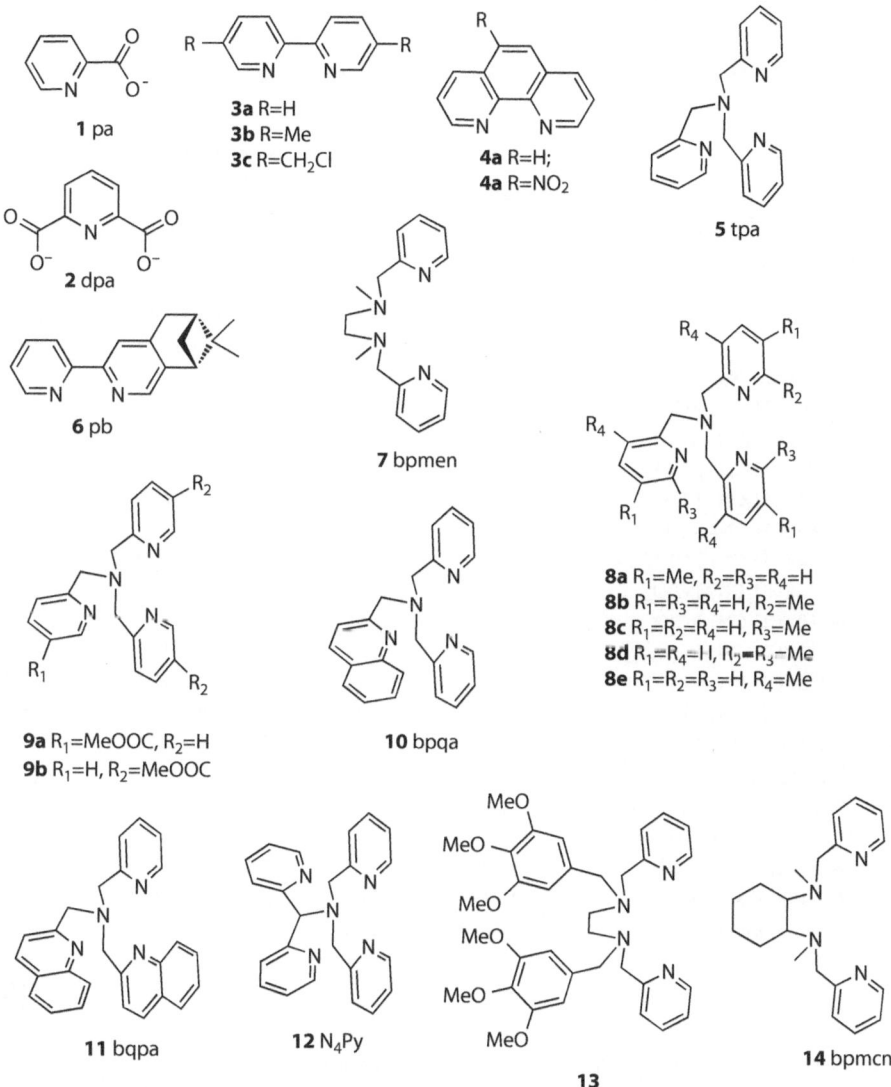

FIGURE 6.1 Polydentate ligands exploited in iron-catalyzed C–H oxidations with H_2O_2.

was observed with up to 72% RC (Table 6.1). This bears evidence in favor of predominantly metal-based rather than radical oxidation [28].

Que with co-workers reported the stereospecific oxidation of alkanes (*cis-* and *trans*-1,2-dimethylcyclohexanes) in the presence of a non-heme iron catalyst [Fe(L)(CH$_3$CN)$_2$](ClO$_4$)$_2$ (L = **5**, tpa) [29]. The latter demonstrated a >99% retention of stereochemistry (substantially higher than the 48 to 72% documented earlier for bipyridine and phenantroline iron complexes [28]), albeit with very low productivities. TONs for cyclohexane oxidation did not exceed 3.7.

TABLE 6.1

Oxidation of Alkanes with H_2O_2 in CH_3CN in Presence of Iron Complexes

N	Catalyst	Cyclohexane Oxidation			Ref.
		A/K Ratio[a]	TON[b]	RC% (Substrate)[c]	
1	$[Fe_2O(3b)_2(H_2O)_2](ClO_4)_4$	Not reported	14.2	48±5 (*trans*-DMCH)	[28]
2	$[Fe_2O(4a)_2(H_2O)_2](ClO_4)_4$	Not reported	4.5	72±5 (*trans*-DMCH)	[28]
3	$[Fe(5)(CH_3CN)_2](ClO_4)_2$	4.3:1.0	3.7	> 99 (*trans*-DMCH)	
				> 99 (*cis*-DMCH)	[29]
4	$[Fe_2O(6)_2(H_2O)_2](ClO_4)_4$	3:1	4.0	100 (*trans*-DMCH)	
				100 (*cis*-DMCH)	[30]
5	$[Fe(7)(CH_3CN)_2](ClO_4)_2$	8:1	6.3	96 (*cis*-DMCH)	[31]
6	$[Fe(8a)(CH_3CN)_2](ClO_4)_2$	9:1	4.0	100 (*cis*-DMCH)	[31]
8	$[Fe(8b)(CH_3CN)_2](ClO_4)_2$	7:1	4.0	85 (*cis*-DMCH)	[31]
9	$[Fe(8c)(CH_3CN)_2](ClO_4)_2$	2:1	2.9	64 (*cis*-DMCH)	[31]
10	$[Fe(8d)(CH_3CN)_2](ClO_4)_2$	1:1	1.4	54 (*cis*-DMCH)	[31]
11	$[Fe(8e)(CH_3CN)_2](ClO_4)_2$	14:1	4.5	100 (*cis*-DMCH)	[31]
12	$[Fe(9a)(CH_3CN)_2](ClO_4)_2$	5:1	4.0	100 (*cis*-DMCH)	[31]
13	$[Fe(9b)(CH_3CN)_2](ClO_4)_2$	19:1	2.3	100 (*cis*-DMCH)	[31]
14	$[Fe(10)(CH_3CN)_2](ClO_4)_2$	10:1	5.8	89 (*cis*-DMCH)	[31]
15	$[Fe(11)(CH_3CN)_2](ClO_4)_2$	2:1	1.7	74 (*cis*-DMCH)	[31]
16	$[Fe(12)(CH_3CN)_2](ClO_4)_2$	1.4:1.0	3.1	27 (*cis*-DMCH)	[32]
17	$[Fe(13)(CH_3CN)_2](PF_6)_2$	8:1	3.8	82 (*cis*-DMCH)	[35]
18	α-$[Fe^{II}(14)(OSO_2CF_3)_2]$	9:1	5.9	>99 (*cis*-DMCH)	[37]
19	β-$[Fe^{II}(14)(OSO_2CF_3)_2]$	0.9:1.0	1.9	68 (*cis*-DMCH)	[37]
20	$[Fe(tpcaH)(MeCN)_2](ClO_4)_2$	7:1	50	96 (*cis*-DMCH)	[38]
21	$[Fe_2O(tpca)_2(H_2O)_2](ClO_4)_2$	7:1	43	97 (*cis*-DMCH)	[38]
22	$[Fe^{II}(17a)(OSO_2CF_3)_2]$	12.3:1.0	6.5	93 (*cis*-DMCH)	[43]
23	$[Fe^{II}(17b)(OSO_2CF_3)_2]$	10.2:1.0	7.6	94 (*cis*-DMCH)	[43]
24	$[Fe^{II}(18)(OSO_2CF_3)_2]$	Not reported	Not reported	100 (*trans*-DMCH)	
				100 (*cis*-DMCH)	[46]
25	$[Fe(19b)(OTf)_2]$	—[d]	70[d]	99 (*cis*-DMCH)	[47]

[a] Alcohol:ketone ratio.

[b] TON calculated in moles of alcohol plus ketone per mole of catalyst.

[c] RC = retention of configuration of resulting 1-monoalcohol: for *cis*-DMCH, RC = 100% × [((1R,2R)+ (1S,2S)) − ((1R,2S)+(1S,2R))]/[((1R,2R)+(1S,2S))+((1R,2S)+(1S,2R))]. For *trans*-DMCH, RC = 100% × [((1R,2S)+(1S,2R)) − ((1R,2R)+(1S,2S))]/[((1R,2R)+(1S,2S))+((1R,2S)+(1S,2R))].

[d] H_2O_2 added in excess; cyclohexanone was major oxidation product.

Menage and co-workers reported a 100% retention of configuration for the oxidation of *cis*- and *trans*-1,2-dimethylcyclohexanes conducted by a dinuclear catalyst $[Fe_2O(6)_2(H_2O)_2](ClO_4)_4$ with pinene-derived chiral ligand **6** (pb) [30]. They also reported small enantiomeric excesses for the epoxidation of ethylbenzene (7% *ee*) and 1,1-dimethylindane (15% *ee*), although whether the asymmetric induction appeared at the hydroxylation stage or the non-racemic mixture resulted from subsequent kinetic resolution of the resulting alcohol was not determined [30].

Que's group later studied the hydroxylation of *cis*-DMCH with H_2O_2 in the presence of iron complexes with *N*-donor polydentate ligands **7–12** (Table 6.1) [31–34]. Iron catalysts bearing differently substituted ligands of the tpa and bpmen families demonstrated different levels of stereospecificity. The loss of stereospecificity correlated linearly with the increasing amount of ^{18}O incorporation (from added $^{18}O_2$), indicating the competition between the rebound of the nascent alkyl radical onto the metal center and either the epimerization of the tertiary *cis*-1,2-dimethylcyclohexyl radical or trapping of the cyclohexyl radical by O_2. The introduction of substituents at the pyridine rings was presumed to slow the rebound step and thus deteriorate the oxidation stereospecificity [31].

It was hypothesized that the presence of two *cis*-labile coordination sites was essential for metal-based (stereospecific) catalysis [31,35,36]. Otherwise, the iron complexes reacted with H_2O_2 through an outer-sphere electron transfer to yield HO· radicals, and further oxidation was driven by freely diffusing HO· in a non-stereospecific fashion [35]. The crucial role of *cis*-α- coordination topology of the catalyst for achieving good catalytic efficiency and high alcohol and ketone selectivity was disclosed. When both *cis*- coordination isomers (*cis*-α and *cis*-β) of [Fe^{II}**(14)** were prepared independently, the *cis*-α complex demonstrated much higher stereospecificity (>99% retention of configuration in *cis*-1,2-dimethylcyclohexane oxidation) than the *cis*-β complex (68% RC) [37].

cis-α-[Fe (bpmcn) (OTf)₂]   cis-β-[Fe (bpmcn) (OTf)₂]

X⁻=CF₃O₂SO⁻ (OTf⁻)

Shteinman and co-workers reported that iron(II) complexes [Fe(tpcaH)(MeCN)₂] (ClO₄)₂ and [Fe₂O(tpca)₂(H₂O)₂](ClO₄)₂ with polypyridyl-based ligand tpcaH (**15**; Figure 6.2) demonstrated reasonably high efficiency in cyclohexane oxidation with H_2O_2 (approaching 50 turnovers) along with 96 to 97% retention of configuration in the oxidation of *cis*-1,2-dimethylcyclohexanes at high [H_2O_2]:[metal] ratios (140 to 420) [38]. They noticed that at lower [H_2O_2]:[metal] ratios (<10), the stereospecificity increased to 100% [39].

The observed trend was interpreted in terms of two active ferryl intermediates [39]. Nordlander and Shteinman prepared a series of tetranuclear iron(III) complexes. Some demonstrated productivities similar to that of the Fe(tpa)/H₂O₂ system and high stereospecificity in *cis*-1,2-dimethylcyclohexane oxidation (93 to 94% retention of configuration) [40].

Feringa and co-workers synthesized iron(II) complexes [LFe(CH₃CN)₂](ClO₄)₂ (where L = N₃Py-Me and N₃Py-Bn, structures of type **16** in Figure 6.2) and tested their

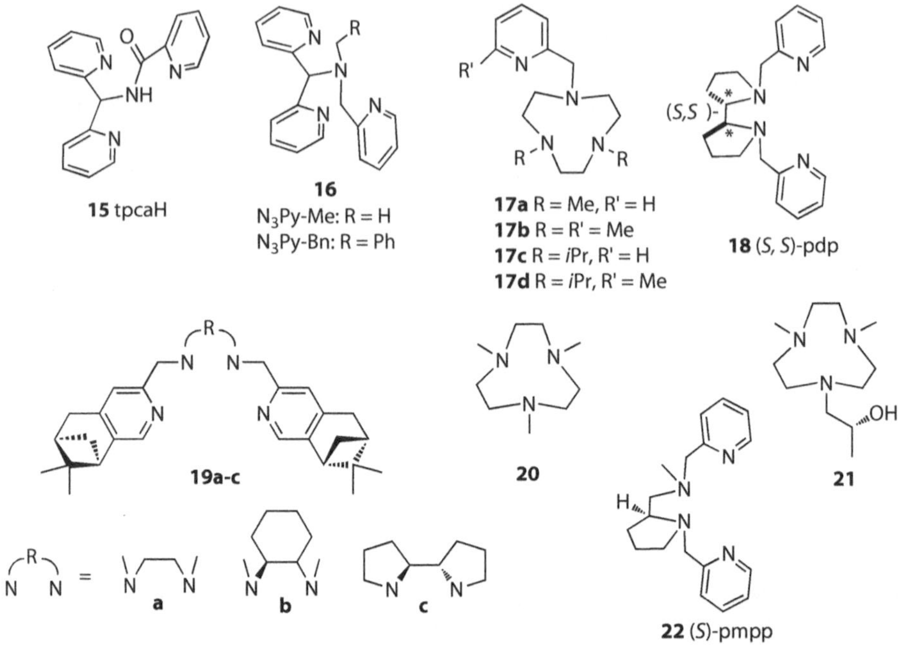

FIGURE 6.2 Polydentate ligands exploited in iron- and manganese-catalyzed C–H oxidations with H_2O_2.

catalytic performance in two different solvents (CH_3CN and acetone). In acetone, the stereospecificities in the oxidation of *cis*- and *trans*-1,2-dimethylcyclohexanes were generally lower and ascribed to the existence of competitive oxidation pathways [41].

Costas and Que with co-workers prepared a series of type [Fe(**17**)(OTf)$_2$] catalysts. One of them (in which ligand **17a**=Me$_2$Py-tacn) was highly stereospecific (94% retention of configuration in *cis*-1,2-dimethylcyclohexane oxidation). The alcohol:ketone ratio of 15 in cyclohexane oxidation was documented [42,43]. In a later account, the authors examined the reactivity of catalyst [Fe^{II}(**17b**)(OSO$_2$CF$_3$)$_2$] and reported a >95% retention of *cis*-DMCH configuration in the course of oxidation [44]. Interestingly, in some cases the catalyst demonstrated a higher selectivity for secondary C–H sites rather than for tertiary C–H sites [44].

A milestone work was contributed by Chen and White who synthesized an iron(II) complex [Fe(**18**)(MeCN)$_2$](SbF$_6$)$_2$ with the (*S*,*S*)-pdp ligand **18** and examined its catalytic properties [45]. In contrast to earlier contributions focused on unraveling the mechanistic features of iron–H_2O_2-based catalyst systems under model reaction conditions, White and Chen explored the synthetic potential of their catalytic system on a range of alkane substrates under practical catalytic conditions [45,46].

Generally, using as much as 15 mol% of the catalyst, 1.2 equivalents of H_2O_2, and 50 mol% of an AcOH additive, up to 90% yield of products (ketones in the case of secondary alkenes and alcohols from tertiary alkenes) was reported. The oxidation of enantiomerically pure substrates proceeded stereospecifically [46] and made it possible to oxidize (+)-artemisinin to optically pure (+)-10β-hydroxyartemisinin in 54% isolated yield [45]. The oxidation of (–)-dihydropleuromutilone proceeded diastereoselectively [46].

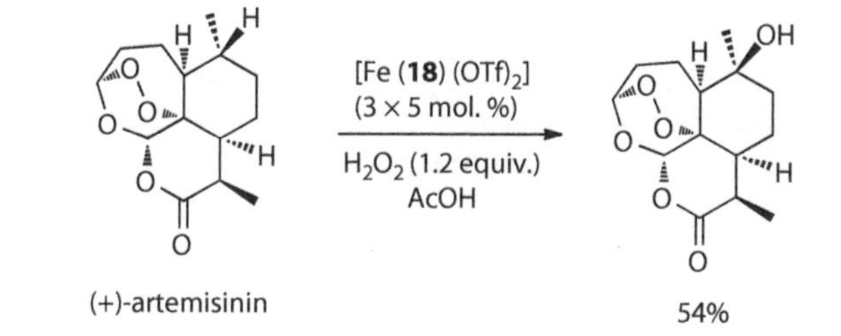

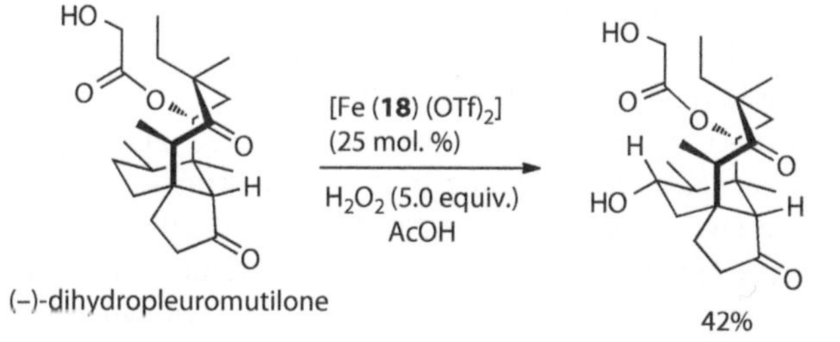

(+)-artemisinin

[Fe (**18**) (OTf)₂] (3 × 5 mol. %)

H₂O₂ (1.2 equiv.) AcOH

54%

(–)-dihydropleuromutilone

[Fe (**18**) (OTf)₂] (25 mol. %)

H₂O₂ (5.0 equiv.) AcOH

42%

(+)-cedryl acetate

[Fe (**19c**) (OTf)₂] (3 mol. %)

H₂O₂ (2.6 equiv.) AcOH

57%

Costas with co-workers extended the steric hindrance of iron catalysts by attaching bulky pinene groups to positions 4 and 5 of the pyridine rings of tetradentate amino-pyridine ligands and prepared a series of novel iron triflate complexes of type [Fe(**19**) (OTf)₂] (where L = **19**) [47,48]. Complexes were tested at catalytic conditions (1 mol% catalyst load, 1.2 equivalents of H₂O₂ with respect to the substrate, 50 mol% of an AcOH additive) in C–H oxidations of various substrates. [Fe(**19b**)(OTf)₂] emerged as an efficient (57% product yield) and stereospecific (99% *cis*) catalyst of *cis*-1,2-dimeth-ylcyclohexane oxidation [47]. The stereospecific oxidation of (+)-cedryl acetate was accomplished in a 57% yield with catalyst Λ-[Fe(**19c**)(OTf)₂] [48].

MANGANESE SYSTEMS

Although manganese-catalyzed C–H oxidations have been extensively studied in the past 15 years [19–22], the authors in very few cases scrutinized or at least mentioned oxidation stereospecificity. For example, Süss-Fink with co-workers prepared binuclear Mn complexes with Me₃tacn **20** and with chiral ligand **21** and tested them as catalysts in the oxidations of various alkanes [49]. In the course of *cis*-1,2-DMCH oxidation, the catalysts showed no stereoretention; furthermore, a more thermodynamically stable *trans*-alcohol was the major product of these reactions [49].

Costas with co-workers reported low efficiencies of complexes [Mn(**14**)(CF$_3$SO$_3$)$_2$] and [Mn(**17b**)(CF$_3$SO$_3$)$_2$] for the stereospecific oxidation of *cis*-1,2-dimethylcyclo-hexane, such that the catalysts performed ≤8 turnovers [47]. To date, we know of only one example of highly efficient stereospecific C–H oxidation with H$_2$O$_2$ catalyzed by manganese complexes. In particular, complexes [Mn(**14**)(CF$_3$SO$_3$)$_2$], [Mn(**18**)(CF$_3$SO$_3$)$_2$], and [Mn(**22**)(CF$_3$SO$_3$)$_2$] demonstrated unprecedented high efficiency, regioselectivity and stereospecificity [50].

A >99% retention of configuration in the oxidation of *cis*-1,2-dimethylcyclohex-ane and 850 to 970 catalytic turnovers (that allowed catalyst loads as low as 0.1 mol%) were reported for all complexes [50]. Similarly, in the presence of catalyst [Mn(**18**)(CF$_3$SO$_3$)$_2$], (–)-acetoxy-*p*-methane converted to the corresponding alcohol with quantitative stereoretention (Figure 6.3).

To conclude this chapter, we must note that the development of general synthetic methods for direct selective (including stereospecific) functionalization of aliphatic C–H groups is a challenging problem. Such methods will exert broad impacts on chemical synthesis spanning all areas of synthetic organic chemistry, including the pharmaceutical and agricultural sectors and the petrochemical industry [51].

Today, we are seeking keys to resolving this problem. Fortunately, the efforts have become less driven by trial and error and more established on solid theoretical bases. In the past 20 years, iron–aminopyridine-based catalysts have demonstrated great potential in C–H oxidations with H$_2$O$_2$. The versatile structure of the tetradentate *N*-donor ligands allow fine tuning of catalytic activities and selectivities by varying both electronic and steric properties of the ligands. While a wide selection of ami-nopyridine and related ligands designed for iron-catalyzed oxidations is available, manganese systems remain undeveloped and very challenging because they may perform more efficiently and selectively in certain situations. Hopefully, manganese-based systems can "deliver the goods" in the near future.

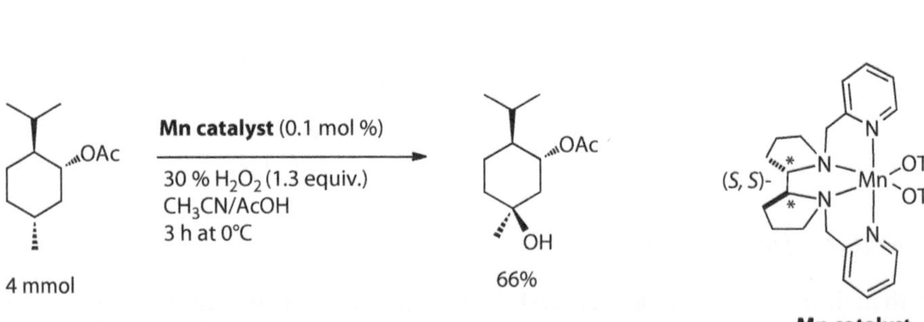

Gram-scale stereospecific hydroxylation of (–)-acetoxy-p-menthane [50].

FIGURE 6.3   Example of preparative scale catalyzed stereospecific hydroxylation.

REFERENCES

1. IUPAC Gold Book. http://goldbook.iupac.org/A00484.html; http://goldbook.iupac.org/S05994.html
2. Shilov, A. E. and Shul'pin, G. B. 1997. Activation of C–H bonds by metal complexes. *Chem. Rev.* 97: 2879–2932.
3. Costas, M., Mehn, M. P., Jensen, M. P. et al. 2004. Dioxygen activation at mononuclear non-heme iron-active sites: enzymes, models, and intermediates. *Chem. Rev.* 104: 939–986.
4. Kryatov, S. V., Rybak-Akimova, E. V., and Schindler, S. 2005. Kinetics and mechanisms of formation and reactivity of non-heme iron–oxygen intermediates. *Chem. Rev.* 105: 2175–2226.
5. Shteinman, A. 2008. Iron oxygenases: structure, mechanism, and modeling. *Russ. Chem. Rev.* 77: 945–966.
6. Hermans, I., Spier, E. S., Neuenschwander, U. et al. 2009. Selective oxidation catalysis: opportunities and challenges. *Top. Catal.* 52: 1162–1174.
7. Berkessel, A. 2006. Diversity-based approaches to selective biomimetic oxidation catalysis. In *Advances in Inorganic Chemistry*, Vol. 58, van Eldik, R. and Reedijk, J., Eds. Amsterdam: Academic Press, pp. 1–28.
8. Tanase, S. and Bowman, E. 2006. Selective conversion of hydrocarbons with H_2O_2 using biomimetic non-heme iron and manganese oxidation catalysts. In *Advances in Inorganic Chemistry*, Vol. 58, van Eldik, R. and Reedijk, J., Eds., Amsterdam: Academic Press, pp. 29–75.
9. Ingold, K. U. and MacFaul, P. A. 2000. Distinguishing biomimetic oxidations from oxidations mediated by freely diffusing radicals. In *Biomimetic Oxidations Catalyzed by Transition Metal Complexes*, Meunier, B., Ed. London: Imperial College Press, pp. 45–89.
10. Costas, M., Chen, K., and Que, Jr., L. 2000. Biomimetic non-heme iron catalysts for alkane hydroxylation. *Coord. Chem. Rev.* 200: 517–544.
11. Shul'pin, G. B. 2002. Metal-catalyzed hydrocarbon oxygenations in solutions: the dramatic role of additives: a review. *J. Mol. Catal. A Chem.* 189: 39–66.
12. Tshuva, E. Y. and Lippard, S. J. 2004. Synthetic models for non-heme carboxylate-bridged di-iron metalloproteins: strategies and tactics. *Chem. Rev.* 104: 987–1012.
13. Korendovych, I. V., Kryatov, S. V., and Rybak-Akimova, E. V. 2007. Dioxygen activation of non-heme iron: insights from rapid kinetic studies. *Acc. Chem. Res.* 40: 510–521.
14. Christmann, M. 2008. Selective oxidation of aliphatic C–H bonds in the synthesis of complex molecules. *Angew. Chem. Int. Ed.* 47: 2740–2742.
15. Comba, P. and Rajaraman, G. 2008. Epoxidation and 1,2-dihydroxylation of alkenes by a non-heme iron model system: DFT supports the mechanism proposed by experiment. *Inorg. Chem.* 47: 78–93.
16. Que, Jr., L. and Tolman, W. B. 2008. Biologically inspired oxidation catalysis. *Nature* 455: 333–340.
17. Shul'pin, G. B. 2009. Hydrocarbon oxygenations with peroxides catalyzed by metal compounds. *Mini-Rev. Org. Chem.* 6: 95–104.
18. Sun, C. L., Li, B. J., and Shi, Z. J. 2011. Direct C–H transformation via iron catalysis. *Chem. Rev.* 111: 1293–1314.
19. Talsi, E. P. and Bryliakov, K. P. 2012. Chemo- and stereoselective C–H oxidations and epoxidations/cis-dihydroxylations with H_2O_2, catalyzed by non-heme iron and manganese complexes. *Coord. Chem. Rev.* 256: 1418–1434.
20. Lyakin, O. Y., Ottenbacher, R. V., Bryliakov et al. 2013. Active species of non-heme – iron- and manganese-catalyzed oxidations. *Top. Catal.* 56: 939–949.

21. Company, A., Lloret, J., Gómez, L. et al. 2012. Alkane C–H oxygenation catalyzed by transition metal complexes. In *Alkane C–H Activation by Single-Site Metal Catalysis*, Pérez, P. J., Ed. Dordrecht: Springer, pp. 143–228.

22. Garcia-Bosch, I., Prat, I., Ribas, X. et al. 2012. Bioinspired oxidations catalyzed by non-heme iron and manganese complexes. In *Innovative Catalysis in Organic Synthesis*, Andersson, P. G., Ed. Darmstadt: Wiley-VCH, pp. 27–46.

23. Shul'pin G. B. 2013. C–H functionalization: by thoroughly tuning ligands at a metal ion, a chemist can greatly enhance catalyst's activity and selectivity (a review). *Dalton Trans.* 42: 12794–12818.

24. Sheu, C., Reichert, S. A., Cofré, P. et al. 1990. Iron-induced activation of hydrogen peroxide for direct ketonization of methylenic carbon c-C_6H_{12} → c-C_6H_{10}(O) and the dioxygenation of acetylenes and arylolefins. *J. Am. Chem. Soc.* 112: 1936–1942.

25. Buxton, G. V., Greenstock, C. L., and Helman, W. P. et al. 1988. Critical review of rate constants for reactions of hydrated electrons, hydrogen atoms and hydroxyl radicals ($\cdot$OH/$\cdot$O$^-$ in aqueous solution). *J. Phys. Chem. Ref. Data* 17: 513–586.

26. Miyajima, S. and Simamura, O. 1975. Stereochemistry of autoxidation of methylcyclohexanes. *Bull. Chem. Soc. Jpn.* 48: 526–530.

27. Gritsenko, O. N., Nesterenko, G. N., and Shteinman, A. A. 1995. Effect of substituents in the ligand on oxidation of alkanes catalyzed by binuclear oxo-bridged iron complexes. *Russ. Chem. Bull.* 44: 2415–2418.

28. Kulikova, V. S., Gritsenko, O. N., and Shteinman, A. A. 1996. Molecular mechanism of alkane oxidation involving binuclear iron complexes. *Mendeleev Commun.* 6: 119–120.

29. Kim, C., Chen, K., Kim, J. et al. 1997. Stereospecific alkane hydroxylation with H_2O_2 catalyzed by an iron(II)–tris(2-pyridylmethyl)amine complex. *J. Am. Chem. Soc.* 119: 5964–5965.

30. Mekmouche, Y., Duboc-Toia, C., Ménage, S. et al. 2000. Hydroxylation of alkanes catalysed by a chiral μ-oxo diferric complex: a metal-based mechanism. *J. Mol. Catal. A Chem.* 156: 85–89.

31. Chen, K. and Que, Jr., L. 2001. Stereospecific alkane hydroxylation by non-heme iron catalysts: mechanistic evidence for an Fe^V=O active species. *J. Am. Chem. Soc.* 123: 6327–6337.

32. Roelfes, G., Lubben, M., Hage, R. et al. 2000. Catalytic oxidation with a non-heme iron complex that generates a low-spin Fe^{III}OOH intermediate. *Chem. Eur. J.* 6: 2152–2159.

33. Chen K. and Que, Jr., L. 1999. *cis*-Dihydroxylation of olefins by a non-heme iron catalyst: a functional model for Rieske dioxygenases. *Angew. Chem. Int. Ed.* 38: 2227–2229.

34. Chen K., Costas, M., and Que, Jr., L. 2002. Spin-state tuning of non-heme iron-catalyzed hydrocarbon oxidations: participation of Fe^{III}–OOH and Fe^V = O intermediates. *Dalton Trans.* 672–679.

35. Mekmouche, Y., Ménage, S., Toia-Duboc, C. et al. 2001. H_2O_2-dependent Fe-catalyzed oxidations: control of the active species. *Angew. Chem. Int. Ed.* 40: 949–952.

36. Mekmouche, Y., Ménage, S., Pécaut, J. et al. 2004. Mechanistic tuning of hydrocarbon oxidations with H_2O_2, catalyzed by hexacoordinate ferrous complexes. *Eur. J. Inorg. Chem.* 3163–3171.

37. Costas, M. and Que, Jr., L. 2002. Ligand topology tuning of iron-catalyzed hydrocarbon oxidations. *Angew. Chem. Int. Ed.* 41: 2179–2181.

38. Gutkina, E. A., Rubtsova, T. B., and Shteinman, A. A. 2003. Synthesis and catalytic activity of the Fe(II) and Fe(III) complexes with a new polydentate ligand containing an amide donor. *Kinet. Catal.* 44: 106–111.

39. Turitsyna, E. A., Gritsenko, O. N., and Shteinman, A. A. 2007. Effect of hydrogen peroxide concentration in stereospecific oxidation of alkanes by models of non-heme oxygenases. *Kinet. Catal.* 48: 53–59.

40. Gutkina, E. A., Trukhan, V. M., Pierpont, C. G. et al. 2006. Tetranuclear iron(III) complexes of an octadentate pyridine–carboxylate ligand and their catalytic activities in alkane oxidation by hydrogen peroxide. *Dalton Trans.* 492–501.

41. Klopstra, M., Roelfes, G., Hage, R. et al. 2004. Non-heme iron complexes for stereoselective oxidation: tuning of selectivity in dihydroxylation using different solvents. *Eur. J. Inorg. Chem.* 846–856.

42. Company, A., Gómez, L., Güell, M. et al. Alkane hydroxylation by a non-heme iron catalyst that challenges the heme paradigm for oxygenase action. *J. Am. Chem. Soc.* 129: 15766–15767.

43. Company, A., Gómez, L., Fontrodona, X. et al. 2008. A novel platform for modeling oxidative catalysis in non-heme iron oxygenases with unprecedented efficiency. *Chem. Eur. J.* 14: 5727–5731.

44. Prat, I., Gómez, L., Canta, M. et al. 2013. An iron catalyst for oxidation of alkyl C–H bonds showing enhanced selectivity for methylenic sites. *Chem. Eur. J.* 19: 1908–1913.

45. Chen, M. S. and White, M. K. 2007. A predictably selective aliphatic C–H oxidation reaction for complex molecule synthesis. *Science* 318: 783–787.

46. Chen, M. S. and White, M. K. 2010. Combined effects on selectivity in Fe-catalyzed methylene oxidation. *Science* 327: 566–571.

47. Gómez, L., Garcia-Bosch, I., Company, A. et al. 2009. Stereospecific C–H oxidation with H$_2$O$_2$ catalyzed by a chemically robust site-isolated iron catalyst. *Angew. Chem. Int. Ed.* 48: 5720–5723.

48. Gómez, L., Canta, M., Font, D. et al. 2013. Regioselective oxidation of non-activated alkyl C–H groups using highly structured non-heme iron catalysts. *J. Org. Chem.* 78: 1421–1433.

49. Romakh, V. B., Rherrien, B., Süss-Fink, G. et al. 2007. Dinuclear manganese complexes containing chiral 1,4,7-triazacyclononane-derived ligands and their catalytic potential for the oxidation of olefins, alkanes, and alcohols. *Inorg. Chem.* 46: 1315–1331.

50. Ottenbacher, R. V., Samsonenko, D. G., Talsi, E. P. et al. 2012. Highly efficient, regioselective, and stereospecific oxidation of aliphatic C–H groups with H$_2$O$_2$, catalyzed by aminopyridine–manganese complexes. *Org. Lett.* 14: 4310–4313.

51. Gunnoe, T. B. 2012. Alkane C–H activation by single-site metal catalysis. In *Alkane C–H Activation by Single-Site Metal Catalysis*, Pérez, P. J., Ed. Dordrecht: Springer, pp. 1–15.

7 Active Species and Mechanisms of Non-Heme Fe- and Mn-Catalyzed Oxidations

The mechanisms of catalytic oxidations in the presence of non-heme Fe and Mn complexes have been thoroughly investigated in the past 20 years. It would appear that detailed examinations of the nature, activity, and stability of true catalytically active species would provide clues to our understanding of the overall reaction mechanisms. However, few studies have focused on the active species of oxidations with H_2O_2 catalyzed by manganese complexes with aminopyridine and related ligands.

In contrast, key intermediates of catalytic oxidations (epoxidations, 1,2-dihydroxylations, C–H oxidations) in the presence of aminopyridine–iron complexes have been studied extensively. The interest in iron-aminopyridine systems arises because they are regarded as the best models of non-heme iron-containing enzymes [1–3]. In this chapter, currently available mechanistic data on the studies of non-heme iron-based catalyst systems for hydrocarbon oxidations with H_2O_2 are surveyed briefly.

In most cases this is done without detailed discussions of minor mechanistic details that seem redundant in this synthetically focused book. In particular, we have restricted the discussion to only the true transition metal-mediated mechanisms operating in highly chemo-, regio- and stereoselective oxidations. Possible minimally selective side reactions such as free radical chain processes or those initiated by iron(IV) complexes are not covered, although their existence may in some cases reduce the outcome of the target product.

On the basis of extensive studies, it is generally accepted the oxoiron(V) intermediates play the key role in chemo- and stereoselective oxidations catalyzed by non-heme iron complexes. Herewith, we mostly focus on publications supporting this mechanism. The available mechanistic data are presented in chronological order, followed by a brief overview of related manganese systems. Recent data bear evidence that in the latter, similar oxomanganese(V) active species are the most likely key intermediates of chemo- and stereoselective oxidations.

IRON SYSTEMS

In an early work on $[Fe^{III}(tpa)Cl_2](ClO_4)$-catalyzed oxidation of cyclohexane with *m*-chloroperoxybenzoic acid and *tert*-butyl hydroperoxide in acetonitrile, Que with co-workers suggested the formation of high-valence oxoiron intermediates [4]. They predicted that the basicity of the tpa ligand (1) may help stabilize the putative $Fe^V=O$ reaction intermediate formed via the heterolytic pathway. Formation of similar oxoiron(V) species was suspected in the $[(tpa)_2Fe_2O](ClO_4)_4/H_2O_2$ [5] and $[(tpa)FeCl_2](ClO_4)/t\text{-BuOOH}$ systems [6].

$$Fe^{III}(tpa) + ROOH \longrightarrow Fe^V=O + ROH \qquad \text{(heterolysis)}$$

$$Fe^{III}(tpa) + ROOH \longrightarrow Fe^{IV}=O + RO^{\cdot} + H^+ \qquad \text{(homolysis)}$$

1 tpa   **2 pa**   **4 tfpy**

Barton with co-workers invoked high-valence iron (presumably oxoiron(V)) species when discussing alkane oxidation with H_2O_2 in pyridine and acetic acid in the presence of $Fe(pa)_3$ and $[Fe^{III}(tpa)](ClO_4)_3$ complexes [7]. Fontecave and co-workers hypothesized the formation of dinuclear active species of type **3** comprising the oxoiron(V) moiety in the catalyst system $[Fe_2(bipy)_4(OH_2)_2](ClO_4)_4/H_2O_2$ [8]. On the other hand, Nishida and co-workers assumed that the oxidation of cyclohexane in the presence of dinuclear iron complexes with ligands of types **1**, **4**, and several related ligands, could be conducted by iron(III) hydroperoxo complexes $Fe^{III}\text{-OOH}$ directly [9].

Que and co-authors supposed that a non-heme ligand environment might not support the Fe^V oxidation state, such that an $[(tpa)Fe^{III}(\eta^2\text{-OOH})]^{2+}$ structure for the intermediate of stereospecific alkane hydroxylations by $[(tpa)Fe(CH_3CN)_2](ClO_4)_2/H_2O_2$ catalyst system might be unfavorable [10]. They proposed a similar hydroperoxoiron(III) intermediate in the [(6-Me$_3$-tpa)Fe(CH$_3$CN)$_2$](ClO$_4$)$_2$-catalyzed *cis*-dihydroxylation of olefins with H$_2$O$_2$. The hypothetical [(6-Me$_3$-tpa)Fe(η^2-OOH)]$^{2+}$ species was expected to attack the alkene either directly or via a transient high-valent oxoiron species [11].

5 6-Me$_3$-tpa   6 [N$_4$PyFe (OOH)]$^{2+}$

   Talsi with co-workers detected several low-spin iron(III) hydroperoxo complexes in bipyridine- and phenanthroline-based iron systems by EPR spectroscopy [12]. They found that the decay rates of detected iron(III) intermediates were similar both in the absence and in presence of organic substrates. The conclusion was that the iron(III) hydroperoxo species could not oxidize the substrates directly. In a subsequent work, the same group documented the absence of substrate effect on the decay rate of the [(tpa)Fe^{III}-OOH]$^{2+}$ intermediate [13].

   Que and Feringa with co-workers characterized a related [(N$_4$Py)Fe^{III}OOH]$^{2+}$ intermediate **6** (detected earlier by electrospray ionization mass spectrometry and presumed to be a η^1-hydroperoxo species [14]) by resonance Raman spectroscopy [15–17]. The authors noticed the similarity of the [(N$_4$Py)Fe^{III}OOH]$^{2+}$ species and of the low-spin iron(III) hydroperoxide intermediate called "activated bleomycin" [18].

   Que and Chen studied the [(bpmen)Fe(CH$_3$CN)$_2$](ClO$_4$)$_2$–H$_2$O$_2$ system (featuring the tetradentate bpmen ligand **7** instead of pentadentate N$_4$Py) that catalyzed the stereospecific oxidation of 1,2-dimethylcyclohexane [19]. On the basis of isotopic labeling data (partial incorporation of oxygen from water into the oxidized product was discussed), a formal Fe^V=O intermediate was predicted (see scheme, where ^{16}O atoms are represented in bold and ^{18}O atoms are in normal font) [19]. The key role of iron(III) peroxo species was excluded since the latter could not exchange its oxygens with those of added $H_2^{18}O$. In turn, the iron(IV) oxo intermediate was reasonably excluded since its formation (via hemolytic cleavage of the O-O bond) should be accompanied by the generation of HO· radicals, capable of non-stereospecific oxygenation of *cis*-1,2-dimethylcyclohexane, which was inconsistent with the observed high oxidation stereospecificity (100% retention of *cis*-configuration).

Chen and Que studied the mechanisms of stereospecific alkane hydroxylation by a series of non-heme iron catalysts. On the basis of isotopic labeling studies, the authors found that in the presence of the $[Fe(TPA)(CH_3CN)_2](ClO_4)_2$ catalyst, the resulting alcohol product contained both oxygen atoms from H_2O_2 and from added H_2O [20]. To rationalize those observations, an original exchange mechanism invoking the hypothetical oxoiron(V) intermediate was proposed.

In a subsequent communication, Costas and Que proposed a general reaction mechanism for alkane hydroxylations and alkene epoxidations/*cis*-dihydroxylations over non-heme iron catalysts featuring two *cis*-α labile sites ($[Fe(TPA)(CH_3CN)_2]$ $(ClO_4)_2$ in particular) [21]. Interestingly, for iron-based catalysts with *cis*-β topology (like ($[Fe(6\text{-}Me_3\text{-}TPA)(CH_3CN)_2](ClO_4)_2$), a different mechanism was predicted, without oxygen exchange with labeled water. The partial incorporation of ^{18}O from added $^{18}O_2$ was reported [21].

The authors noticed the low-spin state of the hydroperoxoiron(III) complex (direct precursor of the oxygen-transferring oxoiron(V) intermediate) is a crucial factor for weakening the O-O bond and activating it for cleavage [22]. In turn, the high-spin iron(III) hydroperoxo species were proposed to require further activation provided by isomerization from the $Fe^{III}-\eta^1$-OOH intermediate to the $Fe^{III}-\eta^2$-OOH one that then reacted with the olefin to form the *cis*-diol with both oxygen atoms stemming from hydrogen peroxide [23,24].

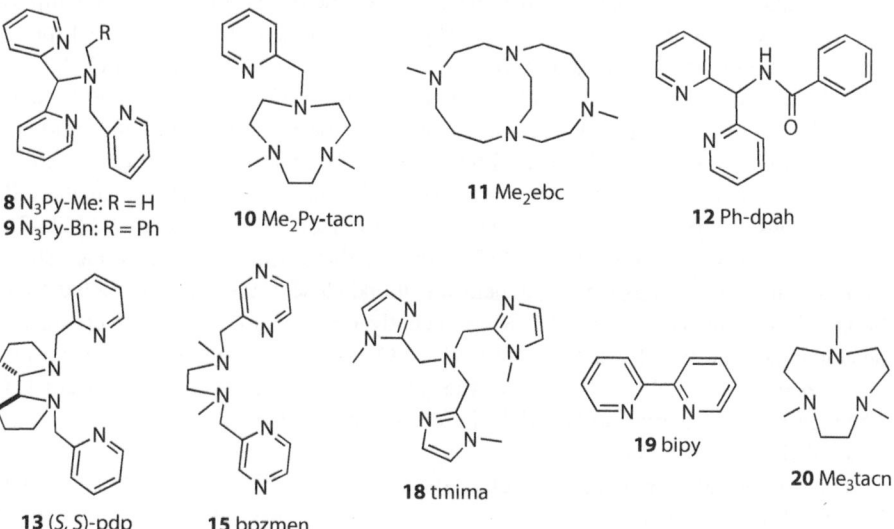

Generation of high-spin peroxoiron(III) intermediates could be favored by the introduction of more than one 6-Me substituent at the tpa ligands (e.g., 6-Me_3-tpa ligand **5**) [23–25]. Formation of similar oxoiron(V) active species was invoked in related systems based on iron complexes with tetradentate N_3Py-Me (**8**; Figure 7.1) and N_3Py-Bn ligands (**9**) [26], bpmen (**7**) [27], Me_2Py-tacn (**10**) [28-30], and Me_2ebc (**11**) [31].

We should note that the concepts mentioned above are valid only for a limited number of extensively studied catalysts and cannot be extended automatically to a broad range of non-heme iron systems. Indeed, a different reaction pathway was assumed for *cis*-dihydroxylation by the $[(Ph\text{-}dpah)_2Fe^{II}](OTf)_2$–$H_2O_2$ system (for the structure of Ph-dpah (**12**); see Figure 7.1 in which the presence of three vacant coordination sites at the iron center in the active species was presumed [32].

8 N_3Py-Me: R = H
9 N_3Py-Bn: R = Ph

10 Me_2Py-tacn

11 Me_2ebc

12 Ph-dpah

13 (*S, S*)-pdp

15 bpzmen

18 tmima

19 bipy

20 Me_3tacn

FIGURE 7.1 Some tetradentate ligands for iron- and manganese-based catalyst systems.

The factors that govern the epoxide–*cis*-1,2-diol oxidation selectivities in the [(tpa)Fe^{II}] and related aminopyridine-based catalysts were discussed in Refs. 22–25, 30, and 33–35. The effect of solvent on the reaction was considered; on the basis of indirect data, Que with co-authors proposed that in acetone, the intermediates in the [(tpa)Fe(CH$_3$CN)$_2$](ClO$_4$)$_2$/H$_2$O$_2$ system could differ from those in acetonitrile, although the reaction products in both solvents were similar [36].

To explain the observed increase of epoxide formation selectivity in the presence of added carboxylic (acetic) acid in non-heme iron systems, Que and co-workers suggested a carboxylic acid-assisted reaction pathway consistent with the formation of a Fe^V=O oxidant [33,37,38]. Indirect evidence in favor of this mechanism was reported by Bryliakov and Talsi with co-authors who found that chalcone epoxidation enantioselectivity in the presence of iron catalyst [(**13**)Fe(OTf)$_2$] increased with raising bulkiness of the carboxylic acid, thus confirming the presence of carboxylic maiety in the structures of the active species [39].

The above conclusions about reaction mechanisms were based mostly on indirect (kinetic and isotopic, computational, and product analysis) data, whereas the reliable detection and identification of the elusive Fe^V=O intermediate were not achieved until recently. In 2007, a model low-spin d^3 oxoiron(V) complex [(taml)Fe^V=O]$^-$ (**14**; Figure 7.2) was synthesized and characterized spectroscopically (EPR, UV-Vis, Mössbauer, and x-ray absorption spectroscopy) at −60°C [40,41].

Talsi and co-workers developed a procedure for the preparation and low-temperature EPR spectroscopic monitoring of unstable low-spin ($S = 1/2$) oxoiron(V) species in various non-heme iron systems (based on tpa (**1**), bpmen (**7**), bpzmen (**15**), (S,S)-pdp (**13**) and other structurally related ligands) with H$_2$O$_2$, *t*-BuOOH, and peracids as oxidants [39,42,43]. The active species were temperature-unstable and gradually decayed already at −70 to −80°C and had similar rhombic EPR spectra (g_1~ 2.6-2.7, g_2~ 2.4, g_3~ 1.5-1.8). Their decay rates were affected by cyclohexene in the EPR sample, thus confirming their key role in catalytic hydrocarbon oxidations [42,43].

For the proposed [(bpmen)Fe^V=O(X)] intermediate, the observed (under steady state conditions) cyclohexene oxide yield was found close to the value predicted on the basis of kinetic measurements [42]. More recently, Cronin and Costas with co-workers detected the catalytically active {[(**10**)Fe^V=O(OH)](OTf)}$^+$ species **16** by cryospray mass spectrometry (above −40°C) in the [(**10**)Fe(OTf)$_2$]/H$_2$O$_2$ catalyst system [44]. The identity of the active oxidant {[(**10**)Fe^V=O(OH)](OTf)}$^+$ was confirmed by isotopic (^{18}O) labeling experiments.

The authors also managed to detect the direct precursors of the diol products, the complexed glycolate species {[(**10**)Fe^{III}(HOCH(R)CH(R′)O)](OTf)}$^+$. In a subsequent joint work, Talsi and Costas with co-workers detected the same (low-spin d^3 [(**10**)Fe^V=O(OH)]$^{2+}$) active species by EPR spectroscopy (g_1 ~2.66, g_2 ~2.43, g_3 ~1.74)

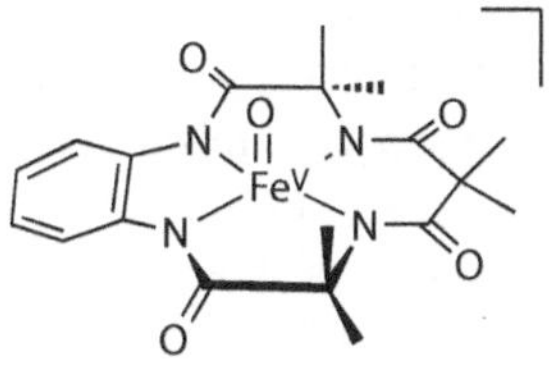

14 [(taml) Fe^V = O]$^-$ 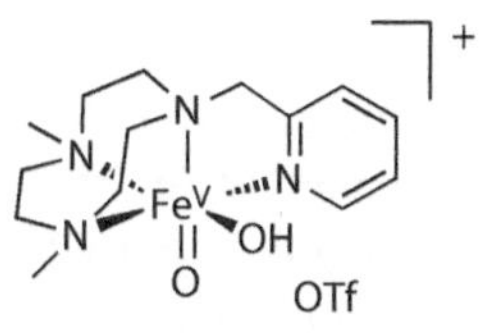   **16** {[(Me$_2$Py-tacn) Fe^V = O (OH)] (OTf)}$^+$

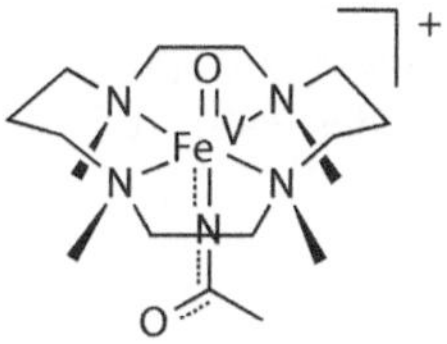

17 [(TMC) Fe^V = O(NC(O)CH$_3$)]$^+$ 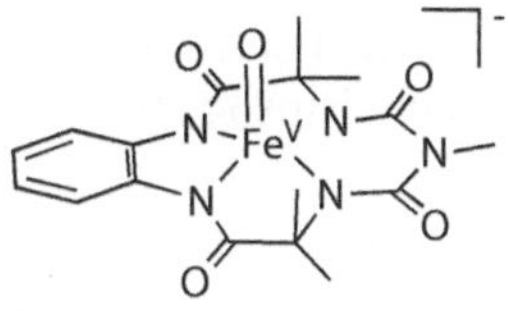   **21**

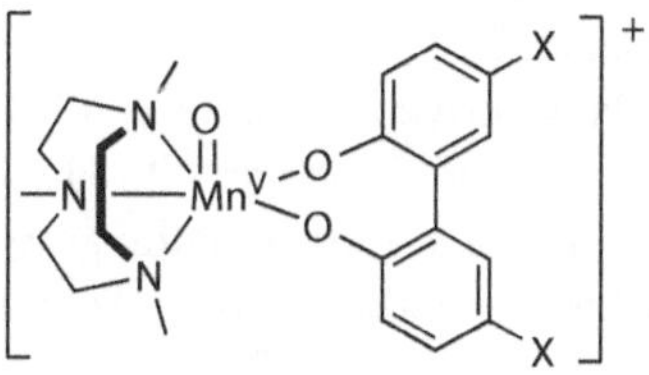

22a X=OMe
22b X=OEt
22c X=H

FIGURE 7.2   Examples of identified oxoiron (V) and oxomanganese (V) complexes.

and confirmed its reactivity to cyclohexene [45]. To date, [(**10**)Fe^V=O(OH)]$^{2+}$ is the only Fe^V=O intermediate with confirmed olefin oxidation activity under catalytic conditions to be detected and identified by two independent physical methods (MS and EPR).

Other examples of recently reported iron(V) complexes are (1) [(TMC) Fe^V=O(NC(O)CH$_3$)]$^+$ (**17**), and its monoprotonated counterpart that were prepared and characterized in situ by various spectroscopic techniques [46], and (2) complex **21**, characterized by UV-Vis, EPR, Mössbauer, and HRMS [47]. The latter was stable at room temperature, yet able to oxidize a series of alkanes having C–H bond dissociation energies up to 99.3 kcal/mol^{-1} (cyclohexane).

MANGANESE SYSTEMS

In contrast to the extensively studied mechanisms of oxidations in the presence of non-heme iron-based catalysts, current knowledge of mechanisms and active species of oxidations with H$_2$O$_2$ catalyzed by non-heme manganese complexes is not supported by extensive and firm experimental background. The formation of a putative

$Mn^V=O$ species was proposed in various catalyst systems based on manganese complexes with polydentate N-donor ligands tmima (**18**), bipy (**19**) [48], and Me_3tacn (**20**) [49–51] but no convincing spectroscopic data were reported in support.

Notable exceptions were reported between 1998 and 2004 when Lindsay Smith with co-workers reported an ESI-MS study of reactive species in the $[(Me_3tacn)_2Mn^{IV}_2(\mu\text{-}O)_3(PF_6)_2]{:}H_2O_2$ system [52]. Formation of oxomanganese(V) species $[(Me_3tacn)Mn^V(O)(5,5'\text{-dialkoxy-2,2'-bisphenolate})]^+$ (**22a,b**), and $[(Me_3tacn) Mn^V(O)(biphenol)]^+$ (**22c**) was presumed on the basis of an ESI-MS data [52–54]. The authors believed that the biphenol ligands stabilized the high-valence state of the oxomanganese species. Using independent techniques, they confirmed the electrophilic nature of the active (presumably oxomanganese(V)) species and proposed general reaction mechanisms for the oxidation of sulfides catalyzed by $Mn\text{-}Me_3$tacn complexes [55].

$$LMn^{IV}Mn^{IV}L \xrightarrow{\ e^-\ } LMn^{III} \xrightarrow{\ H_2O_2\ } LMn^V{=}O \xrightarrow{\ ArSMe\ } LMn^{III}\text{-}O\text{-}\overset{+}{\underset{\backslash}{S}}{}^{Ar} \rightleftharpoons LMn^{III} + O{=}\overset{Ar}{\underset{\backslash}{S}}$$

$L=Me_3$tacn (**20**)

Alternative active species have been considered too. For example, Busch and co-workers proposed a $[(Me_2ebc)Mn^{IV}(O)(OOH)]^+$ active species (see scheme), operating via a Lewis acid oxidant activation pathway on the basis of isotopic labeling studies. The active species did not exchange with $H_2^{18}O$ but ^{18}O incorporation from $^{18}O_2$ to the epoxide, apparently via as radical pathway, was documented) [56]. Feringa with co-workers invoked the formation of binuclear transient active centers in the $[(Me_3tacn)_2Mn^{IV}_2(\mu\text{-}O)_3]^{2+}/H_2O_2$ system [57].

$$LMn^{II}Cl_2 \xrightarrow{\ H_2O_2\ } \left[\begin{array}{c} LMn^{IV}{=}O \\ | \\ OH \end{array} \right]^+ \xrightarrow[-H_2O]{\ H_2O_2\ } \left[\begin{array}{c} LMn^{IV}{=}O \\ | \\ OOH \end{array} \right]^+$$

For manganese-based catalyst systems capable of conducting enantioselective epoxidation of olefins with H_2O_2, sparse mechanistic data have been reported to date. In a combined EPR and enantioselectivity investigation of asymmetric epoxidation of olefins with H_2O_2 in the presence of analogous chiral non-heme iron and manganese complexes, Bryliakov and Talsi with co-workers established a close similarity in the mechanisms of catalytic action of $[((S,S)\text{-pdp})Fe(OTf)_2]/H_2O_2/RCOOH$ and $[((S,S)\text{-pdp})Mn(OTf)_2]/H_2O_2/RCOOH$ catalyst systems in asymmetric olefin epoxidations [39].

Based on similar catalytic behaviors and similar effects of bulkiness of carboxylic acid on the asymmetric induction level, similar oxometal(V) species of the type $[((S,S)\text{-pdp})M^V=O(OOCR)]^{2+}$ (where M = Fe or Mn, R = alkyl) were predicted in both systems [39]. Further support for this conclusion was obtained in a more recent contribution: the same group established the electrophilic nature of the oxygen-transferring species, and showed that the active $[LMn^V=O(OH)]^{2+}$ intermediate could exchange its oxygen with ^{18}O from added $H_2{}^{18}O$, just as the $[LMn^V=O(OH)]^{2+}$ species does [58].

Overall, the nature of catalytically active sites in hydrocarbon oxidations with H_2O_2 catalyzed by non-heme iron and related manganese catalysts has been studied extensively in recent years. Although the problem is not solved yet, convincing independent data support the oxometal(V) hypothesis, especially for the iron systems. Some relevant data are summarized in this chapter. We look forward to further progress, particularly in the detection (and/or synthesis) and characterization of the true high-valence oxygen transferring species.

REFERENCES

1. Costas, M., Chen, K., and Que, Jr., L. 2000. Biomimetic non-heme iron catalysts for alkane hydroxylation. *Coord. Chem. Rev.* 200: 517–544.
2. Costas, M., Mehn, M. P., Jensen, M. P. et al. 2004. Dioxygen activation at mononuclear non-heme iron-active sites: enzymes, models, and intermediates. *Chem. Rev.* 104: 939–986.
3. Que, Jr., L. and Tolman, W. 2008. Biologically inspired oxidation catalysis. *Nature* 455: 333–340.
4. Leising, R. A., Norman, R. E., and Que, Jr., L. 1990. Alkane functionalization by non-porphyrin iron complexes: mechanistic insights. *Inorg. Chem.* 29: 2553–2555.
5. Leising, R. A., Brennan, B. A., Que, Jr., L. et al. 1991. Models for non-heme iron oxygenases: a high-valent iron-oxo intermediate. *J. Am. Chem. Soc.* 113: 3988–3990.
6. Kojima, T., Leising, R. A., Yan, S. et al. 1993. Alkane functionalization at non-heme iron centers: stoichiometric transfer of metal-bound ligands to alkane. *J. Am. Chem. Soc.* 115: 11328–11335.
7. Barton, D. H. R., Beck, A. H., and Taylor, D. K. 1995. The functionalization of saturated hydrocarbons 31. The Fe(pa)$_3$- and Fe(tpa)Cl$_2$.ClO$_4$-catalyzed oxidations of saturated hydrocarbons by hydrogen peroxide: a comparative mechanistic study. *Tetrahedron* 51: 5245–5254.
8. Ménage, S., Vincent, J. M., Lambeaux, C. et al. 1994. μ-Oxo-bridged di-iron(III) complexes and H_2O_2: monooxygenase- and catalase-like activities. *J. Chem. Soc. Dalton Trans.* 2081–2084.

9. Ito, S., Okuno, T., Matsushima, H. et al. 1996. Chemical origin of high activity in oxygenation of cyclohexane by H_2O_2 catalysed by dinuclear iron(III) complexes with amide-containing ligands. *Dalton Trans.* 4479–4484.

10. Kim, C., Chen, K., Kim, J. et al. 1997. Stereospecific alkane hydroxylation with H_2O_2 catalyzed by an iron(II)–tris(2-pyridylmethyl)amine complex. *J. Am. Chem. Soc.* 119: 5964–5965.

11. Chen, K. and Que, Jr., L. 1999. *Cis*-dihydroxylation of olefins by a non-heme iron catalyst: a functional model for Rieske dioxygenases. *Angew. Chem. Int. Ed.* 38: 2227–2229.

12. Sobolev, A. P., Babushkin, D. E., and Talsi, E. P. 2000. Stability and reactivity of low-spin ferric hydroperoxo and alkylperoxo complexes with bipyridine and phenantroline ligands. *J. Mol. Catal. A Chem.* 159: 233–245.

13. Lobanova, M. V., Bryliakov, K. P., Duban et al. 2003. Stability of low-spin ferric hydroperoxo and alkylperoxo complexes with tris(2-pyridylmethyl)amine. *Mendeleev Commun.* 13: 175–177.

14. Lubben, M., Meetsma, A., Wilkinson, E.C. et al. 1995. Non-heme iron centers in oxygen activation: characterization of an iron(III) hydroperoxide intermediate. *Angew. Chem. Int. Ed. Engl.* 34: 1512–1514.

15. Roelfes, G., Lubben, M., Chen, K. et al. 1999. Iron chemistry of a pentadentate ligand that generates a metastable Fe (III)–OOH intermediate. *Inorg. Chem.* 38: 1929–1936.

16. Ho, R. Y. N., Roelfes, G., Feringa, B. L. et al. 1999. Raman evidence for weakened O–O bond in mononuclear low-spin iron (III) hydroperoxides. *J. Am. Chem. Soc.* 121: 264–265.

17. Roelfes, G., Vrajmasu, V., Chen, K. et al. 2003. End-on and side-on peroxo derivatives of non-heme iron complexes with pentadentate ligands: models for putative intermediates in biological iron–dioxygen chemistry. *Inorg. Chem.* 42: 2639–2653.

18. Roelfes, G., Lubben, M., Hage, R. et al. 2000. Catalytic oxidation with a non-heme iron complex that generates a low-spin $Fe^{III}OOH$ intermediate. *Chem. Eur. J.* 6: 2152–2159.

19. Chen, K. and Que, Jr., L. 1999. Evidence for the participation of a high-valent iron–oxo species in stereospecific alkane hydroxylation by a non-heme iron catalyst. *Chem. Commun.* 1375–1376.

20. Chen, K. and Que, Jr., L. 2001. Stereospecific alkane hydroxylation by non-heme iron catalysts: mechanistic evidence for an $Fe^V=O$ active species. *J. Am. Chem. Soc.* 123: 6327–6337.

21. Costas, M. and Que, Jr., L. 2002. Ligand topology tuning of iron-catalyzed hydrocarbon oxidations. *Angew. Chem. Int. Ed.* 41: 2179–2181.

22. Chen, K., Costas, M., Kim, J. et al. 2002. Olefin *cis*-dihydroxylation versus epoxidation by non-heme iron catalysts: two faces of an Fe^{III}–OOH coin. *J. Am. Chem. Soc.* 124: 3026–3035.

23. Chen, K., Costas, M., and Que, Jr., L. 2002. Spin-state tuning of non-heme iron-catalyzed hydrocarbon oxidations: participation of Fe^{III}–OOH and $Fe^V=O$ intermediates. *J. Chem. Soc. Dalton Trans.* 672–679.

24. Fujita, M., Costas, M., and Que, Jr., L. 2003. Iron-catalyzed olefin *cis*-dihydroxylation by H_2O_2: electrophilic versus nucleophilic mechanisms. *J. Am. Chem. Soc.* 125: 9912–9913.

25. Que, Jr., L. 2004. The oxo/peroxo debate: a non-heme iron perspective. *J. Biol. Inorg. Chem.* 9: 684–690.

26. Klopstra, M., Roelfes, G., Hage, R. et al. 2004. Non-heme iron complexes for stereoselective oxidation: tuning of selectivity in dihydroxylation using different solvents. *Eur. J. Inorg. Chem.* 846–856.

27. Mekmouche, Y., Ménage, S., Pécaut, J. et al. 2004. Mechanistic tuning of hydrocarbon oxidations with H_2O_2 catalyzed by hexacoordinate ferrous complexes. *Eur. J. Inorg. Chem.* 3163–3171.

28. Company, A., Gómez, L., Güell, M. et al. 2007. Alkane hydroxylation by a non-heme iron catalyst that challenges the heme paradigm for oxygenase action. *J. Am. Chem. Soc.* 129: 15766–15767.

29. Company, A., Gómez, L., Fontrodona, X. et al. 2008. A novel platform for modeling oxidative catalysis in non-heme iron oxygenases with unprecedented efficiency. *Chem. Eur. J.* 14: 5727–5731.

30. Company, A., Feng, Y., Güell, M. et al. 2009. Olefin-dependent discrimination between two non-heme HO-Fe^V=O tautomeric species in catalytic H$_2$O$_2$ epoxidations. *Chem. Eur. J.* 15: 3359–3362.

31. Feng, T., England, J., and Que, Jr., L. 2011. Iron-catalyzed olefin epoxidation and *cis*-dihydroxylation by tetraalkylcyclam complexes: importance of *cis*-labile sites. *ACS Catal.* 1: 1035–1042.

32. Oldenburg, P. D., Shteinman, A. A., and Que, Jr., L. 2005. Iron-catalyzed olefin *cis*-dihydroxylation using a bio-inspired *N,N,O* ligand. *J. Am. Chem. Soc.* 127: 15672–15673.

33. Bukowski, M. R., Comba, P., Lienke, A. et al. 2006. Catalytic Epoxidation and 1-2 Dihydroxylation of Olefins with Bispidne-Iron(II)/H$_2$O$_2$ systems *Angew. Chem. Int. Ed.* 45: 3446–3449.

34. Bassan, A., Blomberg, M. R. A., Siegbahn, P. E. M. et al. 2005. Two faces of a biomimetic non-heme HO-Fe^V=O oxidant: olefin epoxidation versus *cis*-dihydroxylation. *Angew. Chem. Int. Ed.* 44: 2939–2941.

35. Oldenburg, P. D. and Que, Jr., L. 2006. Bio-inspired non-heme iron catalysts for olefin oxidation. *Catal. Today* 117: 15–21.

36. Mairata i Payeras, A., Ho, R. Y. N., Fujita, M. et al. 2004. The reaction of Fe^{II}(tpa). with H$_2$O$_2$ in acetonitrile and acetone: distinct intermediates and yet similar catalysis. *Chem. Eur. J.* 10: 4944–4953.

37. Mas-Ballesté, R. and Que, Jr., L. 2007. Iron-catalyzed olefin epoxidation in the presence of acetic acid: insights into the nature of the metal-based oxidant. *J. Am. Chem. Soc.* 129:15964-15972.

38. Mas-Ballesté, R., Fujita, M., and Que, Jr., L. 2008. High-valent iron-mediated *cis*-hydroxyacetoxylation of olefins. *Dalton Trans.* 1828–1830.

39. Lyakin, O. Y., Ottenbacher, R. V., Bryliakov, K. P. et al. 2012. Asymmetric epoxidations with H$_2$O$_2$ on Fe– and Mn–aminopyridine catalysts: probing the nature of active species by combined electron paramagnetic resonance and enantioselectivity study. *ACS Catal.* 2: 1196–1202.

40. De Oliveira, F. T., Chanda, A., and Banerjee, D. et al. 2007. Chemical and spectroscopic evidence for an Fe^V–oxo complex. *Science* 315: 835–838.

41. Que, Jr., L. 2007. The road to non-heme oxoferryls and beyond. *Acc. Chem. Res.* 40: 493–500.

42. Lyakin, O. Y., Bryliakov, K. P., Britovsek, G. J. P. et al. 2009. EPR spectroscopic trapping of the active species of non-heme iron-catalyzed oxidation. *J. Am. Chem. Soc.* 131: 10798–10799.

43. Lyakin, O. Y., Bryliakov, and K. P., Talsi, E. P. 2011. EPR, ^{1}H and ^{2}H NMR, and reactivity studies of the iron–oxygen intermediates in bioinspired catalyst systems. *Inorg. Chem.* 50: 5526–5538.

44. Prat, I., Mathieson, J. S., Güell, M. et al. 2011. Observation of Fe(V)=O using variable-temperature mass spectrometry and its enzyme-like C–H and C=C oxidation reactions. *Nat. Chem.* 3: 788–793.

45. Lyakin, O. Y., Prat, I., Bryliakov, K. P. et al. 2012. EPR detection of Fe(V)=O active species in non-heme iron-catalyzed oxidations. *Catal. Commun.* 29: 105–108.

46. Van Heuvelen, C. M., Fiedler, A. T., Shan, X. et al. 2012. One-electron oxidation of an oxoiron(IV) complex to form an O=Fe^V=NR.$^+$ center. *Proc. Natl. Acad. Sci. U.S.A.* 109: 11933–11938.

47. Ghosh, M., Singh, K. K., Pand, C. et al. 2014. Formation of a room temperature-stable Fe^V(O) complex: reactivity toward unactivated C–H bonds. *J. Am. Chem. Soc.* DOI: 10.1021/ja412537m.

48. Fish, R. H., Fong, R. H., Oberhausen, K. J. et al. 1992. Biomimetic oxidation studies 6. Synthetic and mechanistic aspects of manganese cluster-mediated alkane functionalization reactions. *New. J. Chem.* 16: 727–733.

49. Shul'pin, G. B., Nizova, G. V., Kozlov, Y. N. et al 2002. Oxidations by the hydrogen peroxide–manganese(IV) complex–carboxylic acid system. Part 4: Efficient acid–base switching between catalase and oxygenase activities of a dinuclear manganese(IV) complex in the reaction with H$_2$O$_2$ and an alkane. *New J. Chem.* 26: 1238–1245.

50. Woitiski, C. B., Kozlov, Y. N., Mandelli, D. et al. 2004. Oxidations by the hydrogen peroxide–dinuclear manganese(IV) complex–carboxylic acid system. Part 5: Epoxidation of olefins including natural terpenes. *J. Mol. Catal. A Chem.* 222: 103–119.

51. Kozlov, Y. N., Mandelli, D., Woitiski, C. B. et al. 2004. Mechanisms of the oxidations of olefins and alkanes with a H$_2$O$_2$–dimeric Mn(IV) complex–acetic acid system. *Russ. J. Phys. Chem.* 78: 370–374.

52. Gilbert, B. C., Kamp, N. W. J., Lindsay-Smith, J. R. et al. 1998. Electrospray mass spectrometry evidence for an oxo-manganese(V) species generated during the reaction of manganese triazacyclononane complexes with H$_2$O$_2$ and 4-methoxyphenol in aqueous solution. *J. Chem. Soc. Perkin Trans.* 2: 1841–1844.

53. Gilbert, B. C., Lindsay-Smith, J. R., Mairata i Payeras, A. et al. 2004. A mechanistic study of the epoxidation of cinnamic acid by hydrogen peroxide catalysed by manganese–1,4,7-trimethyl-1,4,7-triazacyclononane complexes. *J. Mol. Catal.* 219: 265–272.

54. Gilbert, B. C., Lindsay-Smith, J. R., Mairata i Payeras, A. et al. 2004. Formation and reaction of O=Mn^V species in the oxidation of phenolic substrates with H$_2$O$_2$catalysed by the dinuclear manganese(IV)–1,4,7-trimethyl-1,4,7-triazacyclononane complex Mn$^{IV}_2$(μ-O)$_3$(TMTACN)$_2$.(PF$_6$)$_2$. *Org. Biomol. Chem.* 2: 1176–1180.

55. Lindsay-Smith, J. R., Gilbert, B. C., Mairata i Payeras, A. et al. 2006. Manganese 1,4,7-trimethyl-1,4,7-triazacyclononane complexes: versatile catalysts for the oxidation of organic compounds with hydrogen peroxide. *J. Mol. Catal.* 251: 114–122.

56. Yin, G., Buchalova, M., Danby, A. M. et al. 2006. Olefin epoxidation by the hydrogen peroxide adduct of a novel non-heme manganese(IV) complex: demonstration of oxygen transfer by multiple mechanisms. *Inorg. Chem.* 45: 3467–3474.

57. De Boer, J. W., Browne, W. R., Brinksma, J. et al. 2007. Mechanism of *cis*-dihydroxylation and epoxidation of alkenes by highly H$_2$O$_2$ efficient dinuclear manganese catalysts. *Inorg. Chem.* 46: 6353–6372.

58. Ottenbacher, R. V., Samsonenko, D. G., Talsi et al. 2014. Highly Enantioselective Bioinspired Epoxidation of Electron-Deficient Olefins with H$_2$O$_2$ on Aminopyridine Mn Catalysts. *ACS Catalysis* 4: 1599–1606.

8 Industrial Perspective

GENERAL REMARKS

Catalysis plays a vital role in the chemical, petroleum, agriculture, polymer, electronics, pharmaceutical, and other industries. Over 90% of chemicals originate from catalytic processes [1]. The advancement of methods of asymmetric catalytic synthesis and asymmetric catalytic oxidation in particular is significant for the production of agrochemicals, flavors and fragrances, and pharmaceuticals and vitamins (for which, in many cases, racemic forms are no longer accepted) [2].

This presents several challenges and problems; some arise from the special manufacturing requirements for the products involved; others are due to the nature of the enantioselective catalytic processes. In this chapter, the opportunities and problems associated with the industrial application of stereoselective processes of catalytic oxidation with H_2O_2 and O_2 will be discussed, and some examples of processes that may have industrial potential will be given.

In general, the following considerations are applied when discussing stereoselective catalytic processes (adapted from Ref. [2]):

1. Multifunctional molecules are produced via multistep syntheses (up to 15 steps).
2. The production scale is relatively small, e.g., 1 to 1000 tons per year for pharmaceuticals; the processes usually utilize multipurpose batch equipment.
3. The products are high-value-added molecules, tolerant to high process costs and small production scales.
4. High purity of the final product is essential, i.e., for pharmaceuticals, >99% of the base material containing <10 ppm of metal residue.
5. The development time (or time to market) must be acceptable.

The last factor sometimes assumes primary importance. Indeed, when developing a process for a new chemical entity, the use of time-proven classical technology is very often preferred over more environmentally advantageous processes that need more time to be implemented [3]. In turn, for the so-called second-generation processes (e.g., for chiral switches, generic pharmaceuticals, and non-pharmaceutical fine chemicals), the time factor is not so important and the task is formulated as a design of optimal high performance process rather than a fast adaptation of a competitive process. This may be the case when developing a novel sustainable catalytic asymmetric process to replace an older technology that fails to meet the toughening economic and environmental requirements.

With regard to the viability of a particular novel enantioselective catalyst or a catalytic process, the following major characteristics should be analyzed (adapted from Ref. [2]):

1. The enantioselectivity of the catalytic transformation. For pharmaceuticals, a >99% *ee* is an ideal but rare situation; in most cases >90% *ee* is acceptable provided that further purification is possible.
2. The selectivity: chemoselectivity, regioselectivity, product selectivity, and functional group tolerance.
3. The catalyst productivity. For small-scale products (such as pharmaceuticals), TONs should be >1000 (<0.1 mol% catalyst loads).
4. The catalytic activity must be high enough to ensure complete transformation within acceptable time.
5. Availability and costs of ligands. Chiral ligands prepared from naturally occurring chiral substances (e.g., amino acids, sugars, or terpenes) are usually cheaper than synthetic substances.
6. Availability and costs of starting materials. These can be either too expensive or need additional purification prior to use.
7. Environmental or green chemical characteristics of the planned process, such as atom economy [4] (e.g., for oxidants it is 47% for H_2O_2 and 50% for O_2 [5]), the amount and nature of waste and the E-factor [6] (in case of H_2O_2 and O_2: H_2O, byproduct is not considered a waste and its E-factor is assumed to be zero), the safety and recyclability of solvent, energy efficiency (preferably, a process should be conducted at ambient temperature and pressure).

Although the latter issue does not necessarily favor cost reduction and process simplification, it should be given primary attention because it affects the quality of the environment and human welfare over the long run.

SOME EXAMPLES

In this section, a few catalytic processes (considering the above reservations) will be surveyed. Details are summarized in Table 8.1.

The first process for discussion is the Juliá-Colonna technique developed in early 1980s and aimed at the asymmetric epoxidation of electron-deficient olefins (such as α,β-unsaturated ketones; see also Chapter 5) with hydrogen peroxide in the presence of polypeptide-based catalysts [7–11]. Originally, the process was realized under triphasic conditions (organic solvent–aqueous base–gel-like polypeptide) using H_2O_2 as the oxidant. The main advantages are (1) the use of an environmentally friendly organocatalyst and an inexpensive oxidant and available base (NaOH), (2) the potential to separate and reuse the solid organocatalyst, and (3) the high enantioselectivities (95% *ee* or higher).

Many groups contributed to further development of this synthetic method that became an efficient technology for preparing chalcone-derived epoxides [12,13]. The major technical disadvantages of the original technique (excess oxidant, prolonged

TABLE 8.1
Assessment of Environmentally Sustainable Catalytic Asymmetric Oxidations[a]

Process Characteristics	Juliá-Colonna Epoxidation (Degussa Version) [18][b]	Mn-Catalyzed Epoxidation of Enones [23][b]	Ti-Catalyzed Epoxidation of Olefins [24][e]	Ti-Catalyzed Oxidation of Sulfides [29]	Al-Catalyzed Oxidation of Sulfides [32]
Catalyst	Poly(amino acid)	Mn–aminopyridine complex	Titanium–salalen complex	Titanium–salan complex	Aluminum–salalen complex
Availability of chiral ligands and ligand precursors	Available from chiral pool	Synthetic	Synthetic	Synthetic	Synthetic
Catalyst recyclability and reusability	Recyclable, reusable 25 to 50 times	Non-recyclable	Non-recyclable	Non-recyclable	Non-recyclable
Product	Precursor of bioactive compounds and drugs	Precursors of bioactive compounds and drugs	Model epoxides	Models of precursors of bioactive compounds and drugs	Model sulfoxides
Oxidant	UHP	H_2O_2 (30%)	H_2O_2 (30%)	H_2O_2 (30%)	H_2O_2 (30%)
Reaction time (lowest)	15 min	3 hr	12 hr	16 hr	24 hr
Conversion	>99%	100%	99%	100%	>90%
Enantioselectivity	94% ee	97 to 98%	up to >95%	up to 98.5%	up to 99%
Productivity (TON)	6	500 to >2000	up to 4600	100 (up to 500)	up to 46,000

(continued)

TABLE 8.1 (CONTINUED)
Assessment of Environmentally Sustainable Catalytic Asymmetric Oxidations[a]

Process Characteristics	Juliá-Colonna Epoxidation (Degussa Version) [18][b]	Mn-Catalyzed Epoxidation of Enones [23][b]	Ti-Catalyzed Epoxidation of Olefins [24][e]	Ti-Catalyzed Oxidation of Sulfides [29]	Al-Catalyzed Oxidation of Sulfides [32]
Activity (TOF)	25 hr^{-1}	300 to 500 hr^{-1} or higher	96 hr^{-1}	6 hr^{-1}	up to 1900 hr^{-1}
Atom economy[c]	0.78[b]	0.93[b]	0.89[e]	0.94[f]	0.91[g]
Oxidant excess required	8.8 equivalents	1.3 equivalents	1.01 equivalents	1.2 to 1.3 equivalents	1.1 equivalents
Solvent	THF	CH$_3$CN–2-EHA[d]	CH$_2$Cl$_2$	CH$_2$Cl$_2$	—
Conditions	Ambient	–30°C	Ambient	–10°C	0 to –10°C
Other			N$_2$ atmosphere		Phosphate buffer required

[a] E-factors are not considered; this metric is simple when applied industrially. In a laboratory, the factor is often difficult to estimate since solvent recyclability is not considered.

[b] For chalcone epoxidation.

[c] Calculated as molar mass of target product divided by sum of molar masses of reagents [4].

[d] 2-Ethylhexanoic acid.

[e] For the oxidation of 1,2-dihydronaphthalene.

[f] For the oxidation of 2-naphthyl phenyl sulfide.

[g] For the oxidation of methyl n-octyl sulfide.

catalyst pre-activation period, long reaction times) were successfully overcome. For example, Geller and co-workers at Bayer AG developed a novel high-temperature polypeptide synthesis procedure. Prepared catalysts did not need prolonged pre-activation periods and demonstrated very high activity (in the presence of tetrabutylammonium bromide as a phase transfer agent).

The reaction required only 7 min under triphasic conditions (in the presence of 28.5 equivalents of basic H_2O_2), with retention of quantitative conversion and enantioselectivity up to 97.6% ee [14,15]. The authors also found that the amount of oxidant (H_2O_2) could be reduced from 28.5 to only 1.3 equivalents [16] and developed a scaled-up procedure applicable to hundred-gram substrate loading [17]. The catalyst was re-used without loss of reactivity or enantioselectivity.

Concurrently, researchers at Degussa AG developed a homogeneous version of the Juliá-Colonna process operating in the presence of polyethylene glycol- and polystyrene-bound oligo-L-leucines [18]. The catalyst system demonstrated high catalytic activities that resulted in short reaction times (15 to 60 min). UHP was used as the oxidant and NaOH as the base. A continuously operating reaction setup was developed; the catalyst was retained in the flow reactor via a nanofiltration chemical enzyme membrane. The authors reported catalyst reuse up to 25 residence times, maintaining the conversions over 80% [18].

(L)-Leucine
(polymer-bound)
UHP
NaOH
THF, r. t.

15 min reaction time
>99% conversion
94% ee

$+ H_2O + (NH_2)_2CO$

It is logical to consider transition metal-catalyzed asymmetric epoxidation of electron-deficient olefins in comparison with polypeptide-catalyzed epoxidations. For example, catalysts of the Mn–aminopyridine family reported by various authors (Chapter 2) demonstrated high efficiency and enantioselectivity [19–23]. To date, the original drawbacks such as low turnover (<100), high oxidant excess (up to 6 equivalents), and low enantioselectivity have been solved by fine tuning catalyst structure and reactivity and proper adjustment of reaction conditions [22,23].

At present, chalcone can be oxidized with a practically acceptable 95% optical yield (and quantitative epoxide yield). The catalyst performs at least 1000 turnovers within 3 hr (Mn catalyst 1, Table 8.1) [23]. In test experiments, catalyst 1 demonstrated unprecedented efficiency and enantioselectivity at the same time (7650 TON, 94% ee), albeit at the expense of lower conversion (76.5%) [23].

Mn catalyst 2 appeared even more enantioselective, such that the chalcone epoxide could be synthesized with nearly quantitative yield and 98% ee, albeit with only half efficiency (500 TON) [23]. In contrast to the polypeptide-catalyzed chalcone epoxidation, the Mn–aminopyridine systems are more efficient, enantioselective, and oxidant- and atom-economic. Their disadvantages are the synthetic origin of the catalyst, its non-recyclability, and additional power requirements (for reactor cooling).

A comparative advantage of the Mn system is the broader substrate scope (not only limited to α,β-unsaturated ketones). For example, 2,2-dimethyl-2H-chromene-6-carbonitrile can be oxidized to the corresponding epoxide (a precursor of the levcromakalim potassium channel opener with 96% (catalyst **1**) or 99% (catalyst **2**) *ee* [23].

Prospective titanium–salalen- and titanium–salan-based catalyst systems can be discussed in the context of enantioselective epoxidation of unfunctionalized olefins [24,25]. When used at 1 mol% load, these dinuclear titanium complexes catalyzed the oxidation of a series of conjugated Z-olefins with high yield (up to 99%) and enantioselectivity (up to >99% *ee*) [24]. For at least one of the substrates (1,2-dihydronaphthalene), catalyst loads could be reduced to 0.1 and even to 0.02 mol% without loss of enantioselectivity. The epoxide yield still exceeded 90% within 48 to 72 hr [24].

In a subsequent study, the authors showed that the catalyst also converted non-activated olefins (including terminal ones) to the corresponding epoxides, albeit with lower efficiency (2 mol% catalyst load) and lower yields (19 to 85%) and *ees* (11 to 95%) [25]. For synthetically more readily available titanium–salan complexes (at 5 mol% loads), slightly inferior enantioselectivities (up to 95% *ee* for 1,2-dihydronaphthalene) and efficiencies were documented [26]. Later, it was shown that proper modification of the ligand structures and use of a phosphate buffer led to more robust and enantioselective catalysts [27,28]. We note, however, that the above-mentioned catalyst systems were tested on model substrates and their viabilities for preparative asymmetric epoxidations were not discussed.

Among a variety of catalysts for the enantioselective oxidation of sulfides to sulfoxides, the catalyst systems based on titanium–salan complexes merit discussion. At present, most practical applications exploit various modifications of 30-year-old

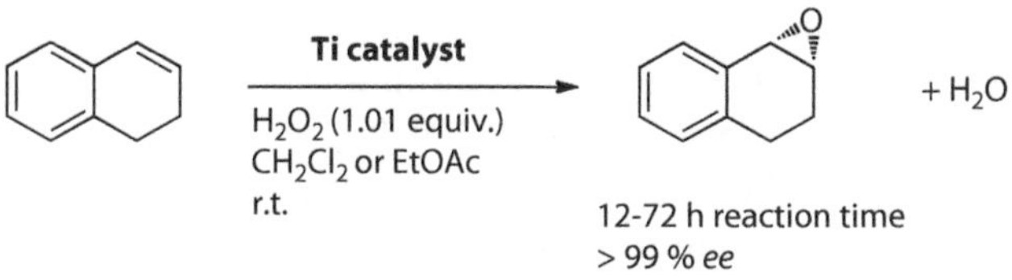

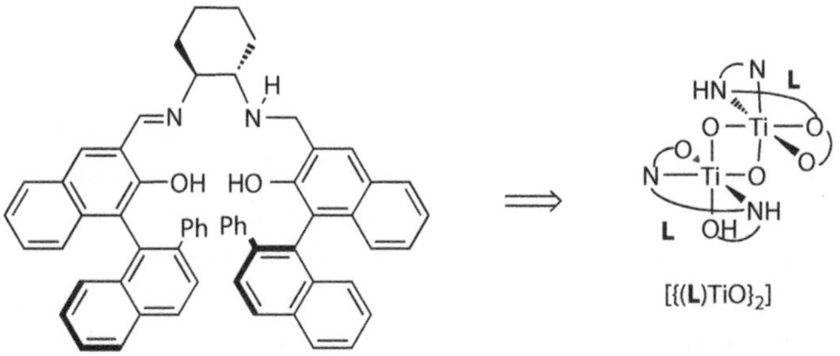

$$[\{(L)TiO\}_2]$$

Ti catalyst

Kagan–Modena systems (based on titanium–dialkyltartrate catalysts) due to their broad substrate scope, tolerance to functional groups, and ease of implementation. Titanium–salan complexes represent a new, "green" generation of titanium-based catalysts.

One reason is the biologically inert nature of titanium and its hydrolysis products that may be important in the production of pharmaceuticals. Furthermore, while most transition metal catalysts discussed in Chapter 3 were tested in the oxidations of small alkyl aryl sulfides (homologues of thioanisole), dinuclear titanium–salan catalysts [29] are capable of conducting the oxidation of bulky aryl benzyl sulfides

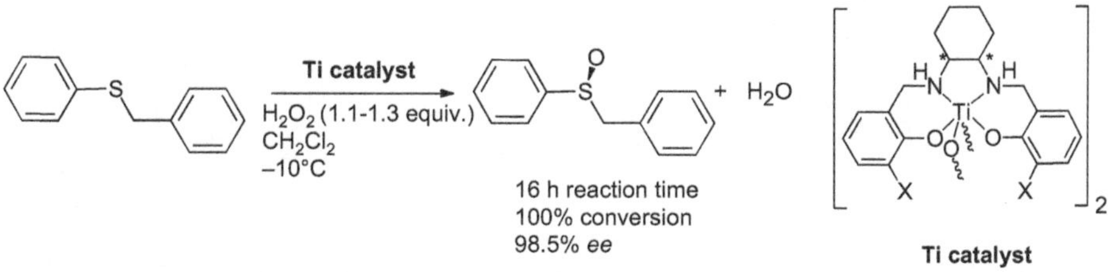

ArSCH$_2$Ph
X=H,CH$_3$, NO$_2$

2-NaphSCH$_2$Ph

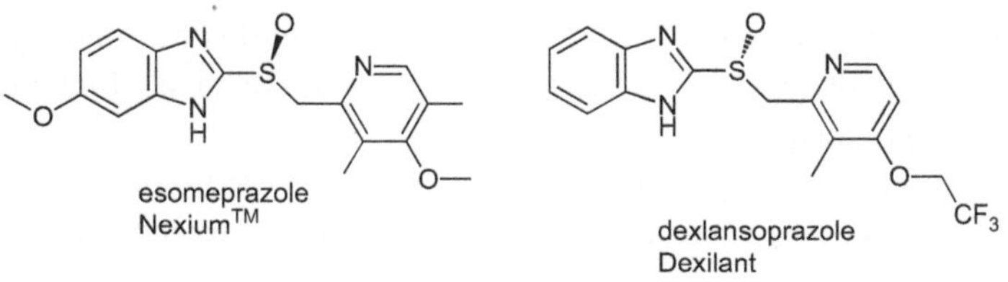

esomeprazole
Nexium™

dexlansoprazole
Dexilant

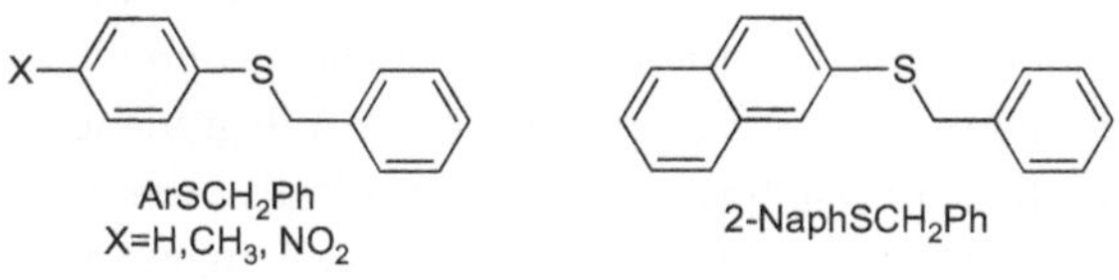

that may be regarded as steric models of precursors of anti-ulcer drugs of the pra-zole family: (S)-omeprazole and (R)-lansoprazole (dexlansoprazole).

The authors tested their catalysts at 1 mol% load; however, the systems can oper-ate at lower loads without loss of enantioselectivity [30]. As compared to the clas-sic Kagan-Modena titanium–dialkyltartrate systems, titanium–salan systems use "green" H_2O_2 as the oxidant (instead of hazardous alkylhydroperoxides). They can perform hundreds of turnovers (versus only 3 to 10), generate much less waste, and do not require aromatic solvents. Their major drawback is the concomitant kinetic resolution occurring in CH_2Cl_2, the solvent of choice.

The kinetic resolution, on the one hand, enhances the optical purity of the sulf-oxide, but it conversely reduces the sulfoxide yield (to 70 to 80% at quantitative con-version) due to overoxidation to sulfone. Another disadvantage is the low turnover frequency that does not substantially exceed that of the titanium–tartrate systems. Nevertheless, although not yet mature enough for technical use, titanium–salan cata-lysts may have the potential to replace the old titanium-tartrate catalysts.

Another potentially attractive catalyst system exploits an Al–salalen combination proposed by Katsuki with co-workers [31,32]. Although tested only on small methyl and ethyl aryl sulfides, the catalysts showed good sulfoxide selectivities (yields of 80 to 95%) and enantioselectivities (97 to 99% ee for least reactive substrates) and extraordinarily high turnover numbers (up to 46,000), thus enabling the authors to reduce the catalyst loads to only 0.01 down to 0.002 mol% [32].

Moreover, the authors found the Al–salalen catalyst capable of operating without a solvent, thus avoiding the need to choose and recycle a proper "green" solvent [33,34]. The Al–salalen catalysts have also demonstrated high chemo- and enanti-oselectivity in the monooxidation of 2-aryl substituted 1,3-dithianes [35].

Disadvantages of this catalyst system are the apparently limited substrate scope, long reaction times, and poor availability of the chiral ligand that must be pre-pared through sophisticated synthetic routes. Another point is that the solvent-free or highly concentrated conditions are not suitable for solid substrates (while most potential precursors of biologically active compounds are relatively high molecular-weight solid compounds).

As for organocatalytic asymmetric sulfoxidations, to date there seem to be no promising examples of organocatalysts that may be adopted by industry in the near future.

OUTLOOK

Asymmetric catalysis is undergoing an evident shift toward more sustainable and "green" catalysts and processes. The area of catalytic asymmetric oxidations with H_2O_2 and O_2 has expanded greatly in the last 25 years, especially since 2000 and many novel metal-based catalyst systems and systems relying on organic catalysts have emerged. It is gratifying to see that the researchers are becoming more interested in sustainable catalyst systems, and the number of those employing environmentally benign dioxygen and hydrogen peroxide as oxidants is steadily expanding.

Titanium- and aluminum–salalen and titanium–salan complexes, manganese–aminopyridine combinations, and polypeptides are only a few authors' choices that demonstrate great synthetic potential as novel environmentally sustainable catalyst systems. While transition metal-catalyzed asymmetric oxidations may be regarded as mature techniques, organocatalytic processes are generally rare and less developed. At present, they cannot complete with metal-based catalysts in the arena of sustainable asymmetric oxidations because metal-based types demonstrate broader substrate scope, higher catalytic efficiencies and reactivities, and in many cases higher enantioselectivities.

In this book, we discussed various catalyst systems and their synthetic potential in well-established asymmetric catalytic transformations such as asymmetric epoxidations, sulfoxidations, non-osmium mediated *cis*-dihydroxylations, Baeyer–Villiger oxidations, kinetic resolution of secondary alcohols, coupling of β-naphthols, and α-hydroxylations of carbonyl compounds. We also emphasized the emerging area of chemoselective stereospecific oxidation of aliphatic C–H groups in the presence of Fe– and Mn–aminopyridine complexes (Chapter 6).

We share the opinion that the development of selective catalysts that functionalize the C–H groups may exert a substantial impact on synthetic chemistry, comparable to those recently crowned by Nobel Prizes in 2001, 2005, and 2010 [36]. On the other hand, the C–H oxidation area (extensively exploiting the biomimetic approach) is one of the most developed in terms of investigating the nature of catalytically active sites operating during oxidation and of complete reaction mechanisms. This prompted us to discuss such studies in detail (Chapter 7). For most other systems and processes, the nature of active species remains either insufficiently explored or arguable; therefore, we did not closely focus on that issue. Nevertheless, the active species and mechanisms are briefly discussed in the text where appropriate.

From an industrial perspective, few of the catalyst systems considered demonstrate truly satisfactory results. In most cases, substantial enhancement of chemo- and enantioselectivity, substrate scope, catalytic efficiency, and reactivity is needed. In addition, such issues as product isolation, catalyst recyclability, ease of implementation, product demand, and cost efficiency must be addressed before such processes can advance to industrial scale (which is, in most cases, not covered in the original research papers). This should not discourage young researchers from joining the field. The road from laboratory to the market is long—perhaps decades. Young researchers have opportunities to participate in development and see advances in the field adopted by industry.

In view of toughening economic and environmental constraints, the demand for efficient and environmentally sustainable chemical processes grows continuously. Society is becoming aware of its responsibility for the quality of life and environmental impacts worldwide. This inspires optimism about the bright future for novel environmentally benign stereoselective catalyst systems and processes. The authors hope that some of catalysts covered in this book may serve as promising alternatives for existing catalysts and progressively replace them in manufacturing processes, ultimately making the chemical industry greener and cleaner. We look forward to future studies that create new opportunities and innovative applications.

REFERENCES

1. Busacca, C. A., Fandrick, D. R., Song, J. J. et al. 2011. The growing impact of catalysis in the pharmaceutical industry. *Adv. Synth. Catal.* 353: 1825–1864.
2. Blaser, H. U., Pugin, B., and Spindler, F. 2005. Progress in enantioselective catalysis assessed from an industrial point of view. *J. Mol. Catal. A. Chem.* 231: 1–20.
3. Sheldon, R. A. 2010. Introduction to green chemistry, organic synthesis, and pharmaceuticals. In *Green Chemistry in the Pharmaceutical Industry*, Dunn, P. J. et al., Eds. Weinheim: Wiley-VCH, pp. 1–20.
4. Trost, B. M. 1991. The atom economy: a search for synthetic efficiency. *Science* 254: 1471–1477.
5. Strukul, G. and Scarso, A. 2013. *In Liquid Phase Oxidation via Heterogeneous Catalysis*. Clerici, M. G. and Kholdeeva, O. A., Eds. Hoboken, NJ: John Wiley & Sons.
6. Sheldon R. A. 1992. Organic synthesis; past, present and future. *Chem. Ind.* 903–906.
7. Juliá, S., Masana, J., and Vega, J. C. 1980. Synthetic enzymes: highly stereoselective epoxidation of chalcone in a triphasic toluene–water–poly(S)-alanine system. *Angew. Chem. Int. Ed. Engl.* 19: 929–931.
8. Juliá, S., Masana, J., Rocas, J. et al. 1982. Synthetic enzymes 2. Catalytic asymmetric epoxidation by means of polyamino-acids in a triphase system. *J. Chem. Soc. Perkin Trans.* 1317–1324.
9. Juliá, S., Masana, J., Rocas, J. et al. 1983. Synthetic enzymes 3. Highly stereoselective epoxidation of chalcones in a triphasic toluene-water-poly(*S*)-alanine. system. *Anal. Quim. Ser. C* 79: 102–104.
10. Colonna, S., Molinari, H., Banfi, S. et al. 1983. Synthetic enzymes 4. Highly enantioselective epoxidation by means of polyamino acids in a triphase system: influence of structural variations within the catalysts. *Tetrahedron* 39: 1635–1641.
11. Banfi, S., Colonna, S., Molinari, H. et al. 1984. Asymmetric epoxidation of electron-poor olefins 5. Influence on stereoselectivity of the structure of poly-α-amino acids used as catalysts. *Tetrahedron* 40: 5207–5211.
12. Berkessel, A. and Gröger, H. 2005. *Asymmetric Organocatalysis*. Weinheim: Wiley-VCH.
13. Gröger, H. 2008. Asymmetric organocatalysis on a technical scale: current status and future challenges. *Ernst Schering Found. Symp. Proc.* 2: 141–158.
14. Geller, T., Gerlach, A., Krüger, C. M. et al. 2004. Novel conditions for the Juliá-Colonna epoxidation reaction providing efficient access to chiral, non-racemic epoxides. *Tetrahedron Lett.* 45: 5065–5067.
15. Geller, T., Krüger, C. M., and Militzer, H. C. 2004. Scoping the triphasic PTC conditions for the Juliá-Colonna epoxidation reaction. *Tetrahedron Lett.* 45: 5069–5071.
16. Geller, T., Gerlach, A., Kruger, C. M. et al. 2006. The Juliá-Colonna epoxidation: access to chiral, non-racemic epoxides. *J. Mol. Catal. A Chem.* 251: 71–77.

17. Gerlach, A. and Geller, T. 2004. Scale-up studies for the asymmetric Juliá-Colonna epoxidation reaction. *Adv. Synth. Catal.* 346: 1247–1249.
18. Tsogoeva, S. B., Woltinger, J., Jost, C. et al. 2002. Juliá-Colonna asymmetric epoxidation in a continuously operated Chemzyme membrane reactor. *Synlett.* 707–709.
19. Wu, M., Wang, B., Wang, S. et al. 2009. Asymmetric epoxidation of olefins with chiral bioinspired manganese complexes. *Org. Lett.* 11: 3622–3625.
20. Ottenbacher, R. V., Bryliakov, K. P., and Talsi, E. P. 2011. Non-heme manganese complexes catalyze asymmetric epoxidation of olefins by peracetic acid and hydrogen peroxide. *Adv. Synth. Catal.* 353: 885–889.
21. Garcia-Bosch, I., Gómez, L., Polo, A. et al. 2012. Stereoselective epoxidation of alkenes with hydrogen peroxide using a bipyrrolidine-based family of manganese complexes. *Adv. Synth Catal.* 354: 65–70.
22. Lyakin, O. Yu., Ottenbacher, R. V., Bryliakov, K. P. et al. 2012. asymmetric epoxidations with H_2O_2 on Fe– and Mn–aminopyridine catalysts: probing the nature of active species by combined electron paramagnetic resonance and enantioselectivity study. *ACS Catal.* 2: 1196–1202.
23. Ottenbacher, R. V., Samsonenko, D. G., Talsi, E. P. et al. 2014. Highly Enantioselective Bioinspired Epoxidation of Electron-Deficient Olefins with H_2O_2 on Aminopyridine Mn Catalyst. *ACS Catalysis* 4: 1599–1606.
24. Matsumoto, K., Sawada, Y., Saito, B. et al. 2005. Construction of pseudo-heterochiral and homochiral di-μ-oxotitanium(Schiff base) dimers and enantioselective epoxidation using aqueous hydrogen peroxide. *Angew. Chem. Int. Ed.* 44: 4935–4939.
25. Sawada, Y., Matsumoto, K., and Katsuki, T. 2007. Titanium-catalyzed asymmetric epoxidation of non-activated olefins with hydrogen peroxide. *Angew. Chem. Int. Ed.* 46: 4559–4561.
26. Sawada, Y., Matsumoto, K., Kondo, S. et al. 2006. Titanium–salan-catalyzed asymmetric epoxidation with aqueous hydrogen peroxide as the oxidant. *Angew. Chem. Int. Ed.* 45: 3478–3480.
27. Matsumoto, K., Sawada, Y., and Katsuki, T. 2006. Catalytic enantioselective epoxidation of unfunctionalized olefins: utility of a Ti(O*i*-Pr)$_4$ salan–H_2O_2 System. *Synlett.* 3545–3547.
28. Shimada, Y., Kondo, S., Ohara, Y. et al. 2007. Titanium-catalyzed asymmetric epoxidation of olefins with aqueous hydrogen peroxide: remarkable effect of phosphate buffer on epoxide yield. *Synlett.* 2445–2447.
29. Bryliakov, K. P. and Talsi, E. P. 2011. Catalytic enantioselective oxidation of bulky alkyl aryl thioethers with H_2O_2 over titanium–salan catalysts. *Eur. J. Org. Chem.* 4693–4698.
30. Bryliakov, K. P. and Talsi, E. P. 2008. Titanium–salen-catalyzed asymmetric oxidation of sulfides and kinetic resolution of sulfoxides with H_2O_2 as the oxidant. *Eur. J. Org. Chem.* 3369–3376.
31. Yamaguchi, T., Matsumoto, K., Saito, B. et al. 2007. Asymmetric oxidation catalysis by a chiral Al(salalen) complex: highly enantioselective oxidation of sulfides with aqueous hydrogen peroxide. *Angew. Chem. Int. Ed.* 46: 4729–4731.
32. Matsumoto, K., Yamaguchi, T., and Katsuki, T. 2008. Asymmetric oxidation of sulfides under solvent-free or highly concentrated conditions. *Chem. Commun.* 1704–1706.
33. Capello, C., Fischer, U., and Hungerbühler, K. 2007. What is a green solvent? A comprehensive framework for the environmental assessment of solvents. *Green Chem.* 9: 927–934.
34. http://www.slideshare.net/RupertStLeger/green-solvent-selection-guide
35. Fujisaki, J., Matsumoto, K., Matsumoto, K. et al. 2011. Catalytic asymmetric oxidation of cyclic dithioacetals: highly diastereo- and enantioselective synthesis of the S-oxides by a chiral aluminum(salalen) complex. *J. Am. Chem. Soc.* 133: 56–61.
36. Gunnoe, T. B. Alkane C–H activation by single-site metal catalysis. In *Alkane C–H Activation by Single-Site Metal Catalysis*, Pérez, P. J., Ed. Dordrecht: Springer, pp. 1–15.

Index